HOW WASHINGTON REALLY WORKS

CHARLES PETERS
Editor-in-Chief
The Washington Monthly

Revised Edition

ADDISON-WESLEY
PUBLISHING COMPANY

Reading, Massachusetts
Menlo Park, California
London • Amsterdam
Don Mills, Ontario
Sydney

Library of Congress Cataloging in Publication Data

Peters, Charles, 1926-
How Washington really works.

Includes index.
1. United States—Politics and government. 2. Power (Social sciences) I. Title.
JK271.P45 1983 320.973 82-20629
ISBN 0-201-14661-4

Revised Edition

ISBN 0-201-14661-4-P

ABCDEFGHIJ-DO-89876543

Dedicated to the Memory of
Richard Rovere

Also by Charles Peters

Blowing the Whistle, with coeditor Taylor Branch

Inside the System,
1st edition, with coeditor Timothy J. Adams
2nd edition, with coeditor John Rothchild
3rd edition, with coeditor James Fallows
4th edition, with coeditor Nicholas Lemann

The Culture of Bureaucracy, with coeditor Michael Nelson

The System, with coeditor James Fallows

FOREWORD

For a great many Americans, Washington no longer seems to work. Like Moslems bowing to Mecca, we keep turning to the capital with hopeful eyes, looking, if not for a miracle, at least for some improvement in our lot in life. When that hope is dashed, we wonder: All that expense, all that calculation, all that exertion of the will—what has it come to? Some disillusioned citizens check all the way out of the great game of politics, having concluded at last that the Washington community is occupied by an essentially foreign power, speaking an alien tongue, addicted to orgies of self-congratulation we have no share in. Politics transforms itself into "us" and "them." "Us" is the Americans, sprawled in bewilderment across a continent. "Them" is Washington, perpetual generator of useless intrusion and insatiable demand.

The pages you are about to read recount tales of that fabulous city, christened for *the* Founding Father, who, when a quarreling Congress could not decide where to put it, set his finger to the map and located our ruling city. Then Washington existed as a village of ambition: in fact a miserable little swamp, in myth a potential Paris. Foreign diplomats had to be paid extra to serve there. Victors in elections prudently left their families at home. No practical thinker expected it to last past more than one or two of Jefferson's nineteen-year generations. But then, like Moses rescued from the bulrushes, Washington survived, flourished, established itself, drawing to it national energies beyond the excitement of New York, the rationality of Philadelphia, the dour disdain of Puritan Boston. For better or worse, Washington, District of Columbia, rose from the ashes of the Revolution as our guiding Phoenix, a tattered bird of passage leading into the future, quintessential representative of the vulgar pioneer spirit.

What then went wrong? There are those who attribute our modern malady to the decline of persons, the two-legged human beings who followed in George Washington's footsteps. His city was progressively invaded by lesser lights. Pointy-headed bureaucrats and Claghorn congressmen, not to mention presidents of the Harding mentality, took over and sullied the springs of political virtue. Smarter tyrants followed and the reaction thereto. The problem of government became the problem of leadership: the search for better men and women to clean up the capitol and get the nation moving again.

Just about the time the current new generation of politically-conscious Americans appeared in the nurseries, John F. Kennedy appeared in The Presidency, a prophet of a higher standard of excellence than had theretofore prevailed. Among those he lured to the nation's capital was one Charles Peters of West Virginia, assigned with consent to find and tell what the Peace Corps was or was not achieving. What Mr. Peters found was mixed; what he learned to tell was plain. The proof of the pudding in American politics had to do, not with intentions, but with results, real changes in the lives of real people. Peters came to Washington by the great circle route—beginning out there where the programs were meant to make a difference, winding up at the center of action where the programs were made. The gap between results and intentions strained his soul, stimulating wonderment.

Peters, like his familial namesake, was a hoper. Perhaps the hope really did reside in the Kennedy thrust: recruiting to government service the best and the brightest of a new generation. Perhaps, in alternative, the essential problem was structural: cutting through the red tape of organizational complexity with the knife of decisive, rational reorganization. But the more Charles Peters experienced of Washington, and the more he contemplated the gap between intention and result, the more it dawned on him that he was dealing with something more fundamental than the circulation of elites or the shuffling of structures. Something more powerful. Something visceral. Something at the heart of the enterprise.

Washington had become a culture. Peters perceived that this Capital of Political Culture owed its continued existence to a set of inherited modes of belief and expectation that gripped the city's practitioners at least as powerfully as did the organization charts. Washington really worked by a set of mores—values thought natural—increasingly divergent from the country's common sense. *That* was the crisis of the twentieth century's waning years: a crisis of vision. Inside the capital city, isolated from the criteria of performance the rest of us had come to take for granted, a peculiar tribal ethic had developed, subject to anthropological analysis, a myopic morality of the salon. The rest of "us" had better understand how that works, because that is where the power is.

For all his skepticism, Charles Peters remains an inveterate hoper. His vision sights forward, toward a humane, decent, practical society where doing the right thing is not the exception but the rule.

I think he would travel with Emerson: "When it gets dark enough you can see the stars." With him, I hope so.

James David Barber
Cullowhee Spring
North Carolina

October, 1979

ACKNOWLEDGMENTS

Regular readers of *The Washington Monthly* will recognize on almost every page my debt to its writers and editors. I also thank Doe Coover, James Fallows, Robert M. Kaus, Michael Kinsley, Nicholas Lemann, Joseph Nocera, and Carol Trueblood for the many helpful suggestions they have made.

I owe special gratitude to my wife, Beth, for her wise counsel, which has been a continuing blessing throughout our life together, and for her dinner table, which for over a year was lost beneath a jumble of documents that were essential to this book but disastrous to civilized dining.

INTRODUCTION

A couple of things should be kept in mind while reading this book. First, I use the word Washington to mean not only the physical place but also our system of politics and government. Second, I'm sure my description of how that system works will strike many readers as highly cynical, and that bothers me a little. I do believe that most of what the government appears to do is make believe carried on for the benefit of those in office, not the rest of us. But I don't believe that it follows that we should abandon hope.

For anybody of my generation, it's necessary only to think back to the early months of our involvement in World War II to realize how dramatically the government's performance can improve. Our triumph at Midway in 1942, one of the greatest naval victories in the history of the world, came just six months after the bombing at Pearl Harbor had demonstrated how abysmally inefficient and oblivious to reality government at its worst can be.

Unfortunately, in recent years our system has too often functioned as it did at Pearl Harbor. That is why I have written this book. I want to see more Midways—in the wars against hunger, disease, ignorance, injustice, and war itself, as well as in the military battles we may be unable to avoid. And I know the Midways will happen only if we learn the lessons of the Pearl Harbors. That task is more difficult today because there is much less dedication than there was in 1942. Certainly, there are dedicated people in government—at the White House, in Congress, in the military, in the foreign and civil services—and they will, I fear, be angered by this book, because they will think I am talking about them. But I'm not talking about them, I'm talking about the majority of people who work for this government, people who seem not to work very hard or very efficiently or to care very much about improving. Some of them, to put it simply, should be fired. Until we have a government that can dismiss the incompetent, we are not going to have effective government. But most are victims of institutional imperatives that tend to inspire the worst in public servants and in the lobbyists and journalists who influence their actions.

These imperatives are usually known or at least dimly perceived by the insider; but even though they diminish his performance, he has little interest in changing them because they also secure his tenure. They are largely unknown by even the well-informed, let alone the average, outsider—a fact that indicts both the journalists and the academics, who are supposed to enlighten the public—and this is why the effort to describe them is such an important part of the pages that follow. They add up to one basic fact: The present system is designed to protect those within it, not to serve those outside. I hope this book will make you want to change it.

CONTENTS

HOW WASHINGTON REALLY WORKS

WELCOME TO THE CLUB
Mouche

LOBBIES

"Dear Dick," the letter began. "As I told you on the telephone, our firm has represented ITT . . . since its incorporation over fifty years ago." The letter was addressed to Richard Kleindienst, then deputy attorney general. It was written by Lawrence Walsh, a former deputy attorney general, a former federal judge, and, when he wrote the letter, a Wall Street lawyer representing ITT. Walsh was asking Kleindienst to do a favor for his client. The favor, which opened the door for ITT to settle an antitrust suit in a way that let the conglomerate keep the assets it wanted most, was granted. When Kleindienst was asked why he went along, he explained, "It was the relationship I had with Judge Walsh, the fact that he was a former deputy attorney general. . . . He wasn't an ordinary attorney as far as I'm concerned."

Kleindienst and Walsh had a special relationship because they were both members of the same "club": present and former high officials of the Justice Department. Kleindienst knew that he too would someday be a former deputy attorney general, and, consciously or unconsciously, he was treating Walsh the way *he* wished to be treated upon his return to private practice. By taking care of ITT's "little problem," Kleindienst added another link to a great chain of favors, a *quid pro quo* that both men understood without having to specify a form for Kleindienst's future *quo*.

Kleindienst's favor to ITT became public—and

controversial—because it became enmeshed in the larger Watergate scandal. But by Washington standards, even by post-Watergate standards, what Kleindienst did was not scandalous at all. After all, he received no money from Walsh. His action was the simple and straightforward result of successful lobbying, and hundreds of similar incidents take place in Washington every business day.

The narrow, legal definition of a lobbyist, as prescribed by the federal lobbying act, is one who "solicits, collects, or receives contributions where one of the main purposes is to influence the passage or defeat of congressional legislation and the intended method of accomplishing the purpose is through conversation with members of Congress." This definition, however, has more loopholes than a spiral notebook. It omits, for example, people who lobby the executive branch; organizations that can show that lobbying is not their main purpose for raising money; individuals or groups who use their own money to influence legislation; and those who don't personally approach members of Congress. Thus, only about 5,000 lobbyists must register under federal law.

But there are really millions of lobbyists. When you write your congressman, you are lobbying, because broadly speaking, lobbying is any attempt to influence the action of a public official. Private individuals and corporations lobby the executive branch as well as Congress. Groups representing various causes or even the "public interest" also lobby both these branches of government. In addition, the branches lobby each other: Congressmen try to influence executive or agency decisions, and the White House lobbies Congress to support its proposals.

The popular conception of a lobbyist is someone who passes money under the table, arranges for clandestine midnight assignations, or holds the threat of blackmail not very high over an official's head. In fact, much of the activity that falls within the broad definition of lobbying is not evil, or even underhanded. When you write a federal official to express an opinion on some public issue, you are trying to make democracy work. You know perfectly well, however, that your letter is not likely to get much attention, so you consider various actions to make sure attention *is* paid. You may be tempted to give a cash bribe, for example. But old Washington hands have developed more subtle lobbying techniques, which are equally effective and have the added advantage of being legal.

The story of Richard Kleindienst and Judge Walsh illustrates the most effective legal way of influencing a public official: having something in common, being a member of the same "club." This is why so many former congressmen work as lobbyists on Capitol Hill.

The mutual membership can even be in an actual Washington club like Burning Tree (golf) or the Metropolitan (lunch). Burning Tree forbids lobbying on its premises, but the friendships formed there have been central to many lobbyists' success. Charls (yes, that's the way he spells it) Walker, perhaps the most successful corporate lobbyist of the 1970s and Ronald Reagan's chief advisor on tax policy during the 1980 campaign, plays a lot of golf at Burning Tree. At the Metropolitan Club in downtown Washington, members bask in the glow conferred by the combination of money, power, and social position held by the people who are present each day in the dining room. Sometimes, merely taking a federal official to lunch there so he can share in the glow for a little while is enough to win his sympathetic attention.

Clubs are just part of a larger social bond that exists everywhere but is especially prevalent in Washington, where private life is so much an extension of professional life. This bond is the "survival network," and it is the key to understanding how Washington really works.

Almost everyone in the government, whether he works on Capitol Hill or in the bureaucracy, is primarily concerned with his own survival. He wants to remain in Washington or in what the city symbolizes—some form of public power. Therefore, from the day he arrives in Washington he is busy building networks of people who will assure his survival in power. The smart lobbyist knows he must build networks not only for himself, but for those officials he tries to influence. Each time the lobbyist meets an official whose help he needs, he tries to let that official know—in the most subtle ways possible—that he can be an important part of that official's survival network.

Suppose you are elected to Congress and are invited to dinner by a clever lobbyist. You will find that as soon as you walk into his living room and the introductions begin, you are meeting one person after another who can be valuable to your career. Because you are also likely to be valuable to them, friendships will develop. And in the long run the friendships formed in these social situations can be a powerful force in decision making. Indeed, there are times when the lobbyist will act more out of loyalty to his network than to his client. He knows that the folks at Mobil or Exxon may forget him one day, but his network won't. This is one of those complex truths about Washington that conventional analysts from either the left or the right rarely see. And it is a truth that helps explain why lobbyists go into lobbying. They are surviving.

Most of the best lobbyists are former high government officials—ex-congressmen, ex-White House staff members, ex-cabinet officers, and ex-assistant secretaries. Usually they enter public service with no thought of later turning it to their own advantage. Most of them probably planned at first to return home after their government service ended. But when actually faced with the prospect of going back, they discover that now Washington is home. Their friends are there, their children are in Washington schools, they own Washington houses. And, above all, there is the sense of excitement, of being at the center of events, that living in the capital confers on its residents. They want to stay in that world, but they have only one really marketable skill that can allow them to remain—their knowledge of government. So they become lobbyists. Their friends who are still in government usually realize that the same thing may eventually happen to them, and they take care to be considerate when the lobbyists come calling.

Although it is a by-product of the survival instinct, this empathy is genuine. It tightens and reinforces the bonds of everyone's network, so that, as Nicholas Lemann observed in *The Washington Monthly,* "although Washington is supposed to be a city where power is carefully balanced between groups with contradictory interests, in fact it's a place with a strong sense of shared enterprise, a place where every person you deal with is someone who is either helping you survive now or might conceivably later on."

The reason that membership in a club is such an effective lobbying tool is that very often the key to effective lobbying is simply *access.* If you are a former congressman, belonging to the club of present and former members gives you an actual physical advantage in the access race: The floor of the House and the Senate and legislators' private dining rooms are open to former members and no other lobbyists.

Why is access so vital? If the other side can't get similar access, a lobbyist's views may be all the official ever hears. Especially on smaller issues, where a decision either way won't rock the ship of state too much, whichever side gets to the congressman usually wins. This also can be true on larger and more controversial issues in non-election years, when officials care less about public opinion. Even when the congressman hears other views, the voice of a friend is likely to stand out in the cacophony of opinions.

More insidiously, the psychology of access plays on the fact that most government officials are basically decent people who want to be nice and want to be liked. Faced with a living, breathing fellow human being who wants something very much, with perhaps only an abstract argument on the other side, the natural reaction is to be obliging. That's why if you are a lobbyist, just getting through to a high official and presenting your case, using facts, figures, and persuasion—no favors involved—gives you a good chance for success. In fact, this is the way most lobbying victories are won.

Between club and cash, there is a wide variety of techniques for gaining access to and influence with the public official whose decisions may affect the lobbyist's company, administration, constituents, or cause. There's the innocent favor, for instance—helping the official's son get a summer job, helping his daughter get into Smith, or introducing him to the social, sports, literary, or entertainment celebrities he happens to admire. Nothing of tangible value is exchanged, yet gratitude is earned.

A story about the late Andy Biemiller, a congressman who became an extremely effective congressional lobbyist for the AFL–CIO, is an excellent example of how effective the innocent favor can be. Biemiller knew that a certain congressman was faced with a serious operation. He knew that George Meany, then head of the AFL–CIO, had faced the same operation and had found an outstanding doctor who successfully performed the difficult surgery. Biemiller referred the congressman to this doctor, who again worked his magic. Meany followed up with a personal call. The result: The congressman, who had voted *against* labor before he became ill, voted *with* labor on an important measure that came up just after he left the hospital.

The insidiousness of the innocent favor is that, like access, it plays on the natural and admirable human desire to be nice. Nobody worth buying can be bought for the price of a lunch or a summer camp session for his child, but it's very hard for a decent person to refuse even to talk ("Just talk, Charlie, that's all I ask") to someone who's done him a favor. It's all the more difficult because that favor was most likely done out of a genuine desire to be nice as well as for the opportunity to get a return favor, and the official will sense this.

Taking an official to lunch or dinner is an innocent favor. But you begin to move away from innocence as the number of meals and the size of the guest list increases. Tip

O'Neill's acceptance of $7,000 worth of parties in his honor from Tongsun Park, the lobbyist for South Korea, was wrong in two ways: First, the monetary value of the parties was far from minimal; second, his acceptance signaled to other congressmen that O'Neill and Park were friends, increasing Park's influence with them.

Other common examples of not-so-innocent favors include free trips to hunting lodges and conventions in places like Honolulu and free flights on private—or air force (depending on who's doing the lobbying)—planes. Sometimes the favor is for a member of the official's family. Ed Gregory, a rich businessman who needed the goodwill of the federal government, arranged for his private airplane to transport Miss Lillian and Billy Carter and Ruth Carter Stapleton on several trips.

Al Ullman, chairman of the powerful House Ways and Means Committee, received the following invitation from Jack Valenti, president of the Motion Picture Association of America (Valenti was once an aide to Lyndon Johnson; the MPA hired him as a favor to LBJ):

"Charles Bludhorn [chairman of Gulf and Western, which owns Paramount Pictures] and I are eager for you and your wife to join us and a few others for what I think will be a spectacular evening. Charlie will 'premiere' the remarkable new film 'Godfather II,' at his home in New York City on Saturday evening, December 7. The evening will begin at 6:30 with cocktails, then the movie and dinner to follow. . . . A Gulfstream II will pick you up and return you."

It is impossible to understand the reality of modern lobbying by looking for an explicit *quid pro quo*. Most likely, Bludhorn and Valenti wanted nothing specific from Ullman at that time. All they may have wanted was Ullman's gratitude for a memorable evening, which would ease their problems of access when they did need him later.

There is another way of obtaining gratitude. Tommy Boggs, a Washington lawyer who lobbies on behalf of such clients as Chrysler and General Motors, did some free lobbying for the Carter administration on the Panama Canal Treaty and SALT II. Such activity, needless to say, created a receptive climate for Boggs when his paying clients needed help at the White House. "I work with him all the time," said Anne Wexler, an assistant to President Carter. "If he comes in on behalf of a client, it's my responsibility to put him [in touch] with someone he needs." Robert Gray, a prominent Republican lobbyist, gives free public relations advice to such Reagan administration figures as Ed Meese, James Watt, Anne Gorsuch, and Larry Speakes. "I always ask," Gray says, "if I can do anything to help."

But these more subtle ways of winning access have not totally displaced the direct cash payment. While bribery is generally frowned on, something close to it is quite common: the paid speaking engagement. Organizations invite a representative or a senator to speak. The members chat with him before he goes on stage, share some refreshments after the speech, and then present him with a check, which is delicately described as an "honorarium." This certainly does not tend to make him hostile to their interests.

In 1977, the Senate voted to lower the limit on income from such speeches from $25,000 to $8,625, effective in 1979. In support of the decrease, Senator Robert Byrd argued that the "honoraria circuit" not only took senators away from their duties in Washington, it also diminished "public confidence in this body" by raising apparent conflicts of interest. He said there was "a threat to the Senate's reputation for integrity when business interests, unions, or other pressure groups affected by legislation pay extremely large sums of money for a short speech by a senator to their group." Senator Byrd was one of six senators who in 1978 earned the maximum allowed under the old law ($25,000). In 1979, he voted to keep the $8,625 limit from becoming effective. He was joined by twenty other senators who had also originally supported it. A study by Common Cause showed that in 1978 these senators had earned an average of almost twice as much as the lower limit.

The study also showed that important members of Senate committees received thousands of dollars from interest groups directly affected by the actions of their committees. Jake Garn, the ranking Republican on the Senate Banking, Housing, and Urban Affairs Committee, received $10,000 for speaking to various banking associations in 1978. Bennett Johnston, chairman of the Energy and Natural Resources subcommittee on Energy Regulation, received $13,750 from power companies and other oil- and gas-related groups. Richard Schweiker, ranking Republican on the Labor and Human Resources subcommittee on Health and Scientific Research, and later Reagan's Secretary of Health and Human Services, received $10,300 from the health care industry.

Next to the paid speaking engagement, the nearest legal equivalent to the straight cash bribe is the campaign contribution. It does not have to be very large. Indeed, the more conscientious the congressman, the more likely he is to be troubled by taking large contributions from someone and voting that person's way on legislation. This explains why, except in the larger states, where the amounts tend to be higher, the accepted price of access is $500 for congressmen and $1,000 for senators—not enough to be suspicious, but enough to be helpful in the campaign.

Lobbyists like Tommy Boggs not only make contributions themselves, they solicit them from others. Boggs is said to have raised funds for more Democratic candidates than anyone in Washington. Robert McCandless, a member of the Democratic National Committee's finance council, said of Boggs, "He has the ability to convince clients and his friends of their need to contribute. . . . You've got to be able to tell your clients that if they are going to do business in this town, they better make certain contributions to the party in power and to kep people on the Hill." The American Medical Association has followed that advice perhaps better than any other lobby. Certainly it spends more. From 1973 through 1978, the AMA and its affiliates poured more than $5 million into congressional campaigns.

For many lobbying groups, the campaign contribution is a straightforward purchase of access rather than a commitment to the candidate or his political philosophy. Contributions are made to important congressmen across the political spectrum, and it is not at all uncommon for contributions to be made to both sides of the same political campaign, just to be safe.

One congressman told Jerry Landauer of the *Wall Street Journal,* "Getting elected to Congress is a painful ordeal. When you come out of the cauldron you're extremely grateful to those who helped you financially or with votes. Certainly the donors want influence, they want to be able to come in any time and have you listen. Of course you'll take a longer look at their problems. If it doesn't violate your principles, you'll try to lean their way, especially on an issue that doesn't involve a lot of other people."

Notice that the congressman referred to those who "helped you financially or with votes." Votes can be appreciated more than any amount of money, especially if they can be produced or denied in significant blocs. This is why lobbies with lots of members, like the AFL–CIO or the National Education Association, wield enormous power.

Corporations could use their stockholders as the AFL–CIO has used its members, but they have seldom done so. On one occasion, however, the stockholders demonstrated the influence they could have. It happened in Maryland in 1978. The Baltimore Gas & Electric Company and the Potomac Electric Power Company urged their 66,000 stockholders to protest a measure before the Maryland legislature. The result: an avalanche of letters and phone calls to the legislators. One delegate, Torrey E. Brown, said, "I had a stockholder walk in my house on Sunday to tell me her stock would be worthless if we passed the bill. I've never had that happen before."

Corporations may not be taking advantage of the full power of their stockholders (AT&T is an exception), but they are taking advantage of legislation enacted in the

1970s permitting managerial employees to form political action committees, through which they can make campaign contributions and serve as campaign workers. These committees were pioneered by the labor movement, but are turning out to be a more powerful tool for the corporations.

Presidents lobby, too, using the promise of votes. A popular president can influence congressmen simply by campaigning for them. Even if he is unpopular he can influence their votes on legislation by taking actions that send money to their states. Thus, when President Carter was trying to get the Panama Canal treaty through the Senate, he helped assure the vote of Senator Dennis DeConcini of Arizona by announcing that the government would buy $250 million worth of copper for the nation's strategic stockpile. Arizona is a big producer of copper. The support of Georgia's Herman Talmadge for the treaty was encouraged by Carter's endorsement of a bill that would help Georgia farmers.

Lobbying groups that are well organized and know when and how to raise their voices can have influence on policymakers far out of proportion to the number of votes they can actually deliver. When the American Road Builders were trying to get Georgia congressmen to oppose a bill that permitted the highway trust fund to be tapped for mass transit, they flew two hundred Georgia highway contractors to Washington to have lunch with their representatives. The result: All but one Georgia legislator voted with the highway lobby.

The American Israel Public Affairs Committee has its 11,000 members computer-listed by congressional district so that they can be mobilized immediately to write their congressmen when issues concerning Israel arise. After a trip to the Middle East a few years ago, Senator Charles Percy, previously considered to be an uncritical supporter of Israel, told a reporter: "Israel and its leadership, for whom I have a high regard, cannot count on the United States in the future just to write a blank check." He said he thought Israel had missed some opportunities to negotiate and called Yasser Arafat "more moderate, relatively speaking, than other extremists."

Columnist Tom Braden obtained a memorandum to Percy written by one of his staff members about what happened during the next week: "We have received 2,200 telegrams and 4,000 letters in response to your Mideast statements. . . . They run 95 percent against. As you might imagine, the majority of hostile mail comes from various Jewish communities in Chicago. They threaten to withhold their votes and support for any future endeavors."

For over a hundred years, the federal government was the biggest employer in the District of Columbia. But in 1979 it was surpassed by the "service" sector, which includes the law and public relations firms and the trade associations where the paid lobbyists work. Their rise reflects the trend toward special-interest politics that has accompanied the decline of the political party. Party loyalty used to be a powerful force in determining legislative action. In his book *Wheeling and Dealing,* Bobby Baker tells of the time in the 1950s when his boss Lyndon Johnson, then minority leader of the Senate, managed to get a unanimous Democratic vote for repeal of the Taft–Hartley Act by appealing to party loyalty. Today, with party loyalty replaced by identification with interest groups, such a vote would be split between pro-labor and pro-business senators. Power has been transferred from party to lobby.

The most startling example of that power is the story of Anna Chennault and the 1968 election. Madame Chennault was a lobbyist for the South Vietnamese government who became convinced, through either express or implied assurances from the Nixon camp, that South Vietnam would do better under Nixon as president than under his Democratic opponent, Hubert Humphrey. In the final days of the campaign, precisely when Humphrey was overtaking Nixon in the polls and a peace agreement between the United States and North Vietnam—an agreement that would surely have accelerated Humphrey's momentum—seemed imminent,

Madame Chennault persuaded the South Vietnamese to torpedo the peace talks. Nixon won the election.

What Madame Chennault did to Hubert Humphrey should teach every politician that if a lobbyist thinks Senator A will do a better job for him than Senator B, he just might decide to arrange for Senator B's defeat. This is the threat that every lobbyist—at least, every lobbyist who commands a lot of money or a lot of votes—holds over the head of every official. This threat is seldom used, however. The lobbyist knows that if he fails to defeat Senator B, the good senator thereafter will actively oppose whatever cause the lobbyist supports. So the best lobbyists make a point of never showing anger when they fail to persuade. Says Representative J. J. Pickle of Tommy Boggs, "I've never seen him put on a fuss when he loses. He lives on for the next day."

Like every really smart lobbyist, Boggs knows the importance of being subtle, keeping a low profile. He avoids publicity because he knows it is the lobbyist's main enemy. Remember the case of Judge Walsh and Richard Kleindienst? Judge Walsh's lobbying for ITT, which perfectly embodied these lessons, worked. But at the same time, ITT had another Washington lobbyist, Dita Beard, who had pledged $200,000 to help pay for the 1972 Republican convention, then planned for San Diego. As she wrote her superiors about this gift, "If it gets too much publicity, you can believe our negotiations with Justice will wind up shot down. [John] Mitchell [then attorney general] is definitely helping but can't let it be known."

Washington columnist Jack Anderson got hold of Mrs. Beard's memo and published it, and that was the end of that lobbying effort. It was Judge Walsh who, in lobbyist parlance, carried ITT's water on this one, using the network instead of cash.

The network is the real secret. When Ellen Proxmire gave a surprise birthday party for her husband, William, when he was chairman of the Senate Banking Committee, guess where she gave it. Why, at Tommy Boggs's house, of course.

THE NEW
NEWS
News
News
NEWS

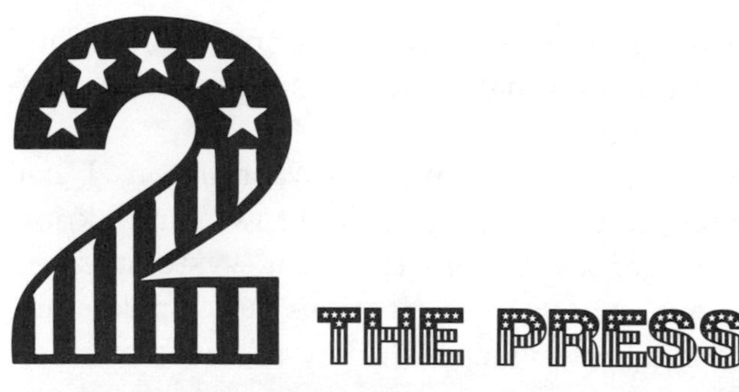

2 THE PRESS

You have been introduced to the first key to understanding Washington, the survival network. The second is "make believe." Washington is like the Winter Palace under Nicholas and Alexandra, where there were constant earnest discussions of the lot of the poor, but the discussions were never accompanied by effective action. In Washington bureaucrats confer, the president proclaims, and the Congress legislates, but the impact on reality is negligible if evident at all. The nation's problems don't disappear, and all the activity that supposedly was dedicated to their solution turns out to have been make believe.

The press, instead of exposing the make believe, is part of the show. It dutifully covers the apparent action—the announcement of programs, the enactment of legislation—rather than finding out how the programs are executed and the legislation is implemented or what the government is *not* doing about crucial problems.

One of the reasons for the persistence of make believe is the press's tradition of the beat system. Reporters are regularly assigned to the White House, to the Congress, to the Pentagon, and to the State Department. They cover official pronouncements but only rarely do they find out whether the new weapons system they write about really works or whether the poor are being hired or the drugs are being tested. Tom Wicker of the *New York Times* explains:

"The problem . . . is that the American press tends to

be an institutionalized press. It covers institutions and processes—anything that has official spokesmen and official visible functions.

"When I was a bureau chief in Washington, I had roughly thirty reporters to deploy around town, and I know how hard it is to get away from the 'beat' system. When they are sent out to cover institutions and spokesmen [which is what the beat system is], they inevitably miss a lot of other things that are happening."

The government likes the beat system as much as the reporters do. Every high official has a press secretary or public information officer, who in turn often has his own platoon—and sometimes an army—of assistants. In 1978 the Department of Health, Education and Welfare published a Directory of Information Sources (the polite term for publicists) that contained 107 names. However, the General Accounting Office said this figure was an understatement. Here, according to the GAO, are the numbers of government employees that were involved in public relations work at some of the more important government offices that year:

Agency	***Employees***
HEW	*338*
Defense	*1,486*
Agriculture	*650*
Treasury	*202*
Congress	*446*
Commerce	*164*
NASA	*208*
Transportation	*117*
HUD	*69*
White House	*85*

All these people make it easy for reporters to get news about government officials. They issue press releases almost every day to make sure the reporters are aware of whatever their bosses have said or done that can be made to look good. They coddle reporters in a warm cocoon of perquisites. Here, as described by congressional correspondent Mick Rood, is a typical day for journalists who cover Congress:

"First stop, the Senate press gallery. It's more spacious than the gallery on the House side of the Capitol. There is time to settle into one of the big, old leather chairs to read a copy or two of the ten daily newspapers supplied us by the Congress. Browsing completed, it's back out past the uniformed guard at the gallery door.

"On your way to some morning coffee, you take the 'Press Only' elevator down to one of the Senate cafeterias. Similar to the 'Senators Only' elevator, the 'Press Only' elevator runs about a floor faster than the public elevators. There is a special section reserved to the press in the cafeteria. Around the corner, the public waits in a long line for a chance to eat breakfast in a Senate restaurant. Here, it's time to engage in colloquy with your colleagues, to discuss the issues of the day and what The Leadership is doing. We could linger, but there is a hearing to cover for the newspaper back home.

"Arriving at the hearing room, you are greeted by another uniformed fellow, who waves you through once you produce your congressional gallery pass. You squeeze by several dozen other people who are standing in line to enter. . . . Once inside, a Senate gallery staff member escorts you to the seat you requested yesterday. Ahead of all others who might be affected by the legislation being considered at the hearing, you are provided copies of testimony by the same gallery staff member. After the hearing, you return to the gallery to bat out a quick story. The typewriter and paper are provided by Congress.

"Lunch hunger calls, but you're out of money. Didn't have time to stop at your bank? No problem. The press gallery superintendent signs off on your personal check. You can cash it downstairs at the Sergeant-at-Arms Bank—along with the legislators and their staffs.

"Off to the House side, where debate has started on the floor on some question or other of great national importance. During debate you may be summoned to the gallery office for telephone calls, which are dutifully logged for you by the gallery staff. The long bank of telephone booths has been provided for in the congressional budget. . . .

"It could be an anniversary, a birthday, or maybe it's Christmas. You've forgotten someone. It's all right. One of the nicest things about serving the people's right to know is the Senate stationery store, where members, staff, *and* reporters can shop at prices below retail.

"As evening falls on Washington, you return to your car, which is parked in the Capitol parking lot. You are among about 180 reporters who share the press slots, bestowed by the House Administration Committee and the Senate Rules Committee." Parking can cost around $70 per month at lots on Capitol Hill, but permit-holding reporters park free.

Reporters assigned to the Defense Department get a lot more than free parking and leather chairs—the Pentagon will fly them across the country and around the world to cover the stories it wants them to print. The White House sometimes even arranges for reporters' families to go along on presidential trips—at one-third to one-half the cost of ordinary fares—to places like Aspen, Key Biscayne, Sea Island, and Santa Barbara. The reporters' part of the bargain is to participate in the make believe that real news is being made in these places rather than just routine statements between rounds of golf.

Because officials are so anxious to get good press, there is often tremendous pressure on the government press agent. Shortly after Robert McKinney became chairman of the Federal Home Loan Bank Board, in the early days of the Carter administration, his public relations officer, Mike Scanlon, arranged a press briefing that put McKinney on the front page of both the *Washington Post* and the *New York Times*. It wasn't long before McKinney came to expect that kind of coverage all the time. When he made a speech in San Francisco that received local publicity but none back East, he fired Scanlon.

Don McClure tells about the time he was serving as public relations officer for the Peace Corps under Sargent Shriver: "One week three magazines—*Newsweek, Look,* and the *Saturday Evening Post*—hit the stands with Peace Corps stories. Shriver wanted to know what had happened with *Time*."

With government press agents operating under this kind of pressure, Washington reporters find stories easy to get. The problem is that they're often *too* easy to get. Here, for example, is the kind of objective news a reporter covering the 94th Congress would have received from the Subcommittee on Government Information and Individual Rights, chaired by Bella Abzug. The titles of the releases:

> "CIA Director Bush Informs Abzug CIA Files to Be Purged"
>
> "Abzug Asks Disclosure of Special Prosecutor Files"
>
> "Abzug Raps Credit Data Abuses"
>
> "Abzug Charges CIA Misled Mail Intercept Victims: Discloses CIA Discovery of New Data on Mail Surveillance"
>
> "Abzug Denounces CIA Intent to Destroy Files"
>
> "Abzug Panel Directs State Department to Bar Race, Religion Questions in Foreign Visas"
>
> "Abzug Panel Reveals Wide Civil Service Use of Subversive Files"
>
> "Abzug Panel Reveals N.Y. Phone Co. Gives Private Unlisted Numbers to 47 Government Agencies, From IRS to Taxi Commission"
>
> "Abzug Panel Discloses Interception of U.S. Cables by British"
>
> "Abzug Releases Justice Department Report on Cointelpro Notification Effort"

People who have reached top levels of government usually have attained their positions at least partly through their skill in handling journalists. They know how to make themselves look good; they also know how to divert attention from the less flattering stories. Reporters who become dependent on these officials, as most do, simply don't get the truth about what's wrong. The most spectacular example of this failure is the case of the White House press corps during the unfolding of the Watergate scandal. Not one of the scores of journalists assigned to full-time coverage of the White House got a major Watergate story. They had been spoon-fed for so long that they lost the habit of independent inquiry. Even when they realized that they had been had, their reaction was not to improve their reporting methods but simply to be rude to the press secretary and to the president.

This rudeness has become another part of the make believe. It makes the members of the White House press corps feel like tough investigative reporters. But all it usually adds up to is a sardonic gloss on a story that the White House wants them to print or broadcast, a veneer of cynicism the public is supposed to interpret as objectivity.

James Fallows writes of his experience with the press when he was head speechwriter during the Carter administration:

"Within the White House, weekly summaries of the President's schedule were prepared; for each day, they listed what the likely 'news event' would be. Under normal circumstances, that prediction almost always came true; if the President was making an announcement about the U.S. Forest Service, the Forest Service would get one day's news—and would not be in the news again until another announcement was planned.

"But after accepting the government's chosen topic, the reporters treat it in their own way, with reflexive cynicism about the administration's plans. The true lesson of Watergate is the value of hard digging, not only into scandal but everywhere else. The *perceived* lesson of Watergate in the White House press room is the Dan Rather lesson, that a surly attitude can take the place of facts or intelligent analysis. More and more often at the President's press conferences, one sees reporters proving their tough-mindedness by asking insulting questions; in the daily briefings with Jody Powell, open snarling became the norm. TV correspondents feel they've paid homage to the shade of Bob Woodward by ending their reports not with intelligent criticism but with a sophomoric twist: 'The administration says its plans will work but the true result is *still to be seen.* Dan Daring, NBC News, the White House.' "

The men who got the Watergate story, Bob Woodward and Carl Bernstein of the *Washington Post*, did not get it by asking questions of the White House press office. One of their sources, Deep Throat, is widely suspected to have been David Gergen, who handled publicity for the Reagan administration but was a low-ranking functionary in the Nixon White House. Those whose identities Woodward and Bernstein later revealed were not high-ranking officials. Instead, they were personal secretaries and middle-level executives like Hugh Sloan, the assistant treasurer of the Committee to Re-elect the President. When Bill Moyers was press secretary to Lyndon Johnson, he said the kind of leaker he feared most was not the cabinet member, who could usually be trusted to guard his comments even after infuriatingly disappointing sessions with the president. Rather, Moyers said, the secretary's special assistant, who heard everything about those sessions once the secretary returned to his office, was much more likely to leak.

The Hugh Sloans and the special assistants are more prone to talk because they are several layers removed from personal loyalty to the president and because they are less skilled than their bosses in fending off the inquiries of the press. There is also another factor that sometimes encourages them to be honest: a loyalty to something they see as being more important than the president. This could be the welfare of the country, or it could simply be the welfare of their agency. The reporter who, next to Woodward and Bernstein, did the best job on Watergate, Sandy Smith of *Time,* got most of his material from middle-level bureaucrats at the FBI, who resented Patrick Gray's attempts to make what they regarded as political use of the bureau. And don't forget James McCord's famous letter to Judge Sirica that broke the Watergate dam. It was motivated by McCord's anger at what Nixon had tried to do to the CIA, where McCord had been a middle-level bureaucrat.

It can be argued that Woodward and Bernstein got the Watergate story precisely because they were not White House correspondents but local reporters for the *Washington Post*'s Metro section. If they had been White House correspondents, they would have asked Ziegler and Nixon what was going on and received the usual runaround. Because they weren't, they had to dig down to the lower-level people who knew the story and were willing to talk.

The people who work on the Metro section of the *Post* often try to follow Woodward's and Bernstein's example and make their reputations by uncovering a scandal. But once a Metro reporter has "made it," he or she gets assigned to a prestigious national affairs beat. Unfortunately, the reporter then tends to become a statesman covering other statesmen. This is exactly what happened to Woodward and Bernstein. For all the inside information in their second book, *The Final Days,* they relied heavily on Alexander Haig and Fred Buzhardt, who were high-level officials during the last year of the Nixon administration. Not surprisingly, the book makes both Haig and Buzhardt appear to be fine fellows, if not outright heroes. In fact, both were far from being innocent bystanders in the conspiracy to obstruct justice that was the main activity of the White House during Nixon's last two years as president.

Which brings us to sources. The leaking of information is a Washington art form practiced by bureaucrats and politicians at all levels. As we have seen, many middle-level bureaucrats leak information to the press either because they resent being manipulated by their bosses or because they are loyal to values higher than personal job security. And occasionally people will leak to the press on a touchy subject to make sure their bosses pay attention to warnings they might otherwise choose to ignore: It's easy to bury a folder of unpleasant statistics in a messy office but hard to overlook them on the front page of the *Washington Post* or the *New York Times.*

In an article on leaking, Joseph Nocera describes several instances in which subordinates, "frustrated at not being able to reach superiors through the chain of command," leaked information to the press:

"In late 1965, for example, people on the China desk at the State Department were worried about U.S. plans to bomb certain North Vietnamese targets close to the China border. They had tried to raise their objections the 'right way'—sending memos to their superiors and hoping that their objections would be heard. But it was clear that they were getting through to no one—certainly not to the three men who counted most: Robert McNamara, Secretary of Defense; Dean Rusk, Secretary of State; and McGeorge Bundy, Johnson's national security advisor. So one day, when Max Frankel, who then covered the State Department for the *New York Times,* was making his rounds, they advised him to take a look at the recent issues of the Peking *Daily,* where warnings about such bombings were being printed. Frankel looked, saw the story, and wrote a front-page piece for the *Times*. Frankel's article, of course, infuriated McNamara, Rusk, and Bundy, but after their wrath had subsided, they held a meeting with the leakers and ended up changing the bombing plans."

Leaking can also be an effective way to push a point of view or to create an atmosphere either favorable or unfavorable to certain policies. It might even take the form of providing incorrect and misleading information to the enemy in wartime. The Allies did this during World War II, by leading the Germans to think they would invade Calais instead of Normandy. The most common reason for leaking, however, is a familiar one: to strengthen the survival network. Officials and reporters build and nurture their networks through mutual favors: "I'll leak this piece of news to you if you'll give my boss front-page coverage on a certain issue. You get a front-page story, and my boss gets publicity." This kind of interdependence weakens a reporter's objectivity and heightens his susceptibility to manipulation.

Syndicated columnists are particularly susceptible to being conned by their important sources. Because they have to turn out several columns a week, they don't have time to piece together stories from dozens of interviews. Therefore, talking to a single top official who supposedly knows the whole story is irresistibly appealing. In addition, readers are impressed when, say, Joseph Kraft recounts his personal interview with a Brezhnev or a Sadat. Getting these exclusive interviews with world leaders also inflates the columnist's own sense of self-importance. Soon it is beneath him even to consider doing any legwork or talking to those more lowly who might tell him something really interesting.

James Reston of the *New York Times* started his career as an outstanding reporter working the periphery, but he is ending it talking to Cy Vances and Alexander Haigs, or sometimes merely observing the movements of the clouds. Evans and Novak's coverage of the Nixon and Ford administrations seemed to be based on one primary source: Melvin Laird, the former congressman and Secretary of Defense, who is turn usually received sympathetic treatment in their column.

The trap, as we saw with Woodward and Bernstein's treatment of Haig and Buzhardt, is that the columnist becomes obligated to give his source favorable treatment. He becomes a prisoner of his source. The more access he is given, the harder it is to criticize. The late Drew Pearson, author of "The Washington Merry-Go-Round," said, "We will give immunity to a very good source as long as the information he offers us is better than what we have on him." According to Mary Anne Dolan of the *Washington Star*, Pearson's successor, Jack Anderson, told a Washington gathering that for some years he had a standing agreement with J. Edgar Hoover to write only nice things about the FBI director in exchange for access to the bureau's files.

When the source is a big shot, the *quid pro quo* is often not hard information but simply the glow of close association with the mighty. Sometimes this can lead to embarrassingly fawning journalism, such as these words written by Kraft about McGeorge Bundy:

"The central fact is that Bundy is the leading candidate, perhaps the only candidate, for the statesman's mantle to emerge in the generation that is coming to power—the generation that reached maturity in the war and post-war period. His capacity to read the riddle of multiple confusion, to consider a wide variety of possibilities, to develop lines of action, to articulate and execute public purposes, to impart quickened energies to men of the highest abilities, seems to me unmatched. To me, anyhow, he seems almost alone among contemporaries, a figure of true consequence, a fit subject for Milton's words:

> A Pillar of State; deep on his
> Front engraven
> Deliberation sat, and publick care;
> And princely counsel in his face."

This was written in 1965, the year Bundy served as a main architect of one of the most disastrous decisions of the Vietnam war—the decision to escalate American troop strength there to over 500,000 men.

The fatuously empty syndicated column represents one unhappy extreme of modern journalism. On the other extreme, the craze for investigative reporting also has its unfortunate aspects. Reporters like Seymour Hersh and Woodward and Bernstein have inspired thousands of imitators. An example of the absurdity that this craze can produce: Recently in San Jose, California, when some poor park attendant apparently neglected to shut off a drain, killing some valuable fish called Koi, the *San Jose Mercury* responded with four-alarm Woodward-and-Bernstein coverage. Here are some of the headlines:

"Valuable Fish Left to Die"

"Parks Chief Alters Story on Fish Kill"

"Parasites Found in Koi Fish"

"Koi Controversy Continues"

"Heated Session on Koi"

"Garza Blasts Parks Chief in Session on Koi Death"

"Fish Death Censure Stalled"

There is another danger of the new wave of investigative reporting: the methods that some reporters use to get information. It's common practice for reporters to adopt poses. For example, they can check on plane reservations, bank loans, or long-distance telephone calls simply by posing as the person being investigated and requesting confirmation of the reservation, an extra copy of the loan agreement, or a duplicate of long-distance bills. Some reporters have gone to great lengths to scoop others on a story. As Timothy Ingram reports: "Al Lewis, the *Washington Post*'s veteran police reporter, for example, was the only newsman inside the Democratic headquarters at the Watergate on the morning the five burglars were arrested. Wearing white socks and looking very much the cop, Lewis simply accompanied the acting police chief past the fifty reporters and cameramen cordoned off from the Watergate complex by the police. Once inside, Lewis took off his jacket, sat down at a desk, and occasionally pecked at a typewriter. He looked for all the world as if he was supposed to be working there. With a phone at his desk, he was able to provide the *Post* with a description of the office floor plan, details about the surgical gloves and lock-picks and jimmies used, and the name of the security guard who foiled the break-in. Lewis sees nothing deceitful in his actions—all he was doing was remaining anonymous. He never *told* anyone he was a policeman, and presumably had anyone asked, he would have disclosed his true identity."

But problems can result from such deceptions. "False premises," Ingram says, "can result in false information. . . . A reporter conceals his identity in order to hear things the source would not intentionally tell the press. But he may also hear things the source would not tell the press because they are untrue: the source may be lying to impress a stranger; the information may be wrong, or couched in terms that are misunderstood; the person may be careless in what he says because he doesn't think he is speaking for the record."

Finally, in a quest for exclusive, behind-the-scenes, on-the-spot news, members of the media may find they are helping to create that news instead of merely reporting it. There is a story that in the mid-1960s, CBS bid over $30,000 for the exclusive film rights to an insurrection in Haiti. But when CBS realized this money was helping the rebel army buy guns and ammunition—that, in effect, CBS was subsidizing the invasion, not just covering it—the network withdrew the offer.

A more common failure of investigative reporting is an article about an alleged abuse in which the significance of the abuse is never made clear. Conflicts of interest are reported as evil per se when they are only potentially evil. (When a reporter reveals that some senator owns stock in a company that does business with the government, the reporter too often stops there, with merely finding the conflict of interest, instead of proceeding to determine whether the conflict has ever actually influenced the senator's official behavior, which is the *real* story.) Worst of all for Washington, the investigative reporter looks for scandalous illegality when he should be looking into why the government doesn't work. What's wrong with government today seldom has much to do with illegality. Occasionally it does, such as when a congressman is on the take from ABSCAM, or when officials in the General Services

Administration are taking bribes from government suppliers. But most of the time the explanation of what is wrong lies in the cultures of the bureaucracy, the Congress, the White House, and the judiciary—that is, in the customs and rituals and pressures that govern life in these institutions. The average reporter is remarkably ignorant of these cultures.

Reporters who understand the culture of the bureaucracy would have known, for example, that one or more tragedies such as My Lai were likely to occur in Vietnam as soon as the Pentagon began publishing body counts of enemy casualties to prove America was winning the war. The reporters would have realized that the pressure for more casualties would lead commanders to find those casualties wherever they could. If reporters understood that mentality, they would not have waited for the story to be revealed by a Vietnam veteran a year after the fact. They might even have prevented it by asking questions and writing stories about the dangerous possibility that the numbers game would lead to shooting animals, women, children—anything that could be counted in the casualty totals.

One more example: In May 1978, the *Washington Post* ran a front-page scandal story telling how one government agency spent money recklessly at the end of its fiscal year in order to use up all its appropriation for the year and avoid risking a budget cut because it hadn't spent all its money. Neither the author of the article nor the editor knew that the end-of-the-fiscal-year spending spree has been a government-wide practice for years. Here was one of the nation's leading newspapers—one that has long had primary responsibility for covering Washington—unaware of an important facet of the culture of the bureaucracy.

Such ignorance of the substance of government is characteristic of today's journalism. Reporters write stories saying that President Reagan is pledged to Cabinet government; but they don't note that so were many of his predecessors, nor do they explore why those other presidents felt they had to abandon that pledge. And they write about Jimmy Carter's firing a group of cabinet members as if it were a national catastrophe. In fact, such firings have happened in the past and are not out of the ordinary.

Ignorance of government may be one reason why reporters focus on elections, the horse-race aspect of politics. The drama of an election can be reported without any knowledge of government. If the candidate wants to avoid the issues, so does the reporter. The result is more make believe—namely, that the candidate who wins the election will make a real difference in how the government operates.

What all this means is that, in addition to looking for the crooks and covering "politics," reporters should be trying to understand the system. This would not only make it possible for them to explain why and how the occasional official crook became crooked, but it could also examine the much more significant problem of why good and decent men produce an inefficient, uncaring, and sometimes evil government.

The real story behind Vietnam, for example, was available to many reporters in Washington. After all, those who had friends in the State Department knew most of them were privately against the war from the day it started until the day it ended. Why didn't these officials say at the State Department what they said at dinner parties? They were afraid of damaging their careers. And why didn't their journalist friends report what was said privately? Because they were in the same survival networks and did not want to hurt the officials' careers. When, in 1968, the *New York Review of Books* sent Mary McCarthy to Vietnam to find out why the war was still going on, she said that she hadn't gone earlier because her husband was in the foreign service and she didn't want to endanger his position. This means she didn't have to go to Vietnam; she could have gotten the story from her husband—and from herself. Why hadn't he resigned in protest and why hadn't she reported his failure? Looked at that way, she might have found the war criminals more understandable.

The influence of the press on government officials cannot be overestimated. This influence can be positive. Think how much chicanery dies on the drawing board when someone says, "We'd better not do that; what if the press finds out?" On the other hand, there are ill effects such as these:

• A public official or a candidate for public office seldom does anything important after 5 P.M. because it won't make the evening news and therefore will not have happened.

• Presidential candidates campaign in the small, out-of-the-way state of New Hampshire because they know reporters are watching.

• Top staff meetings at the White House and in the various agencies and departments are devoted to getting puff pieces written. The puff pieces are then accepted as reality by those who inspired them.

• The transcripts of meetings of the Nuclear Regulatory Commission during the near-disaster of Three-Mile Island indicate that the only time the commissioners devoted sustained attention to one subject was when they debated—sometimes for hours—the wording of a press release. And this during a period when life-and-death decisions affecting hundreds of thousands of people had to be made.

Washington reporters could find out the truth. It doesn't require unusual ability, just the willingness to break free from the conventional beats and go where the *real* action is. But they are too bound up in the make believe system of reporting apparent action, in the ease of being stenographers for government press agents, and in the thrill of rubbing shoulders with the mighty and walking through the White House gate past the admiring eyes of waiting tourists.

William Kovach, the *New York Times*'s bureau chief in Washington, says: "Nothing is more painful than the look on the face of a reporter who has been told that he or she must turn in the White House press pass because someone else needs it more; nor is there any look quite so seraphic as that on the face of one told he or she has just been added to the list of White House press pass holders."

White House reporters will cynically tell you (off the record, of course) all about being spoon-fed by the Jody Powells. They don't say it on the record because they don't want their editors to see it and take them off the White House beat. They know that what they're doing doesn't really count, but the folks back home think it does. It sometimes seems that White House reporters live life for the stories they can tell their friends, stories about Carter and Sadat or Kennedy and his girlfriends. It may be only make believe, but as the press plane takes off for another presidential trip abroad, who cares?

FIRE DEPT.
HOT TUNES
BUREAU EFFICIENCY
What to do when the OMB comes
UNEMPLOYMENT
FIREPERSONS
ROOM 20003

3 THE BUREAUCRACY

There is a permanent government in Washington that consists of people whose power does not depend on election results. It includes the courts, the military, and the foreign services as well as those two unofficial but powerful branches we have just examined, the press and the lobbies. But the largest part of the permanent government is the bureaucracy, which has over two and a half million federal civilian employees. Ten times that number are funded by the taxpayers, either through government contracts, as employees of state, county, and municipal governments, or as members of the military.

With one in ten of our citizens working for government, it is not surprising that the bureaucratic presence is increasingly dominant in our lives. You experience it when you stand in line to get a license, when you try to pry loose your aunt's overdue social security check, when you start to pay your taxes and realize you have to pay even more to hire someone to prepare your return.

Some bureaucrats do essential work. Air traffic controllers, Forest Service fire fighters, nurses at the Veterans Administration hospitals, and researchers seeking a cure for cancer or protecting you from dangerous drugs all perform vital functions that tangibly affect our daily lives.

But for every one of these useful civil servants there is a government employee whose contribution to the public interest is less clear—employees with such titles as Planning Analyst, Schemes Routing Specialist, Manager of Creative Services, Social Priorities Specialist, Suggestions Awards Administrator, Fringe Benefit Specialist, or Confidential Assistant to the Confidential Assistant (all real titles). These bureaucrats have the better-paying jobs and usually work in the District of Columbia. I first encountered them when I moved to Washington from West Virginia in 1961; at that time I was still a fervent believer in the civil service system. In the late 1950s, while working on the staff of my state legislature, I drafted a bill designed to transform a patronage-riddled personnel system into a service based on merit that offered genuine career protection for state employees. My determination to get that bill enacted into law was one of the reasons I ran for a seat in the legislature in the next election, and it was a proud day in my life when the bill, bearing my name, was passed in the following session.

Then I came to Washington. Having seen the evils of too many political hacks, I was now exposed to the evils of too many planning analysts. The old West Virginia system was characterized by too much corruption and too little continuity; the Washington system was characterized by too much defensiveness and too little courage.

Certainly not all bureaucrats are defensive cowards. And of course in many clerical jobs courage or openness is irrelevant. But as people begin to climb the administrative ladder, a dominant personality type emerges—or perhaps certain elements in the personality come to dominate as the people get promoted. Either way, an excess of caution is certainly one of the characteristics. It is ironic that a system invented to protect the courageous and outspoken has attracted people who rarely need protection on those counts. Instead, they are looking for protection against anything that could disturb their quiet but steady progress up the career ladder.

When did the bureaucracy turn into the lethargic, self-protective monster it is today? It became monstrous between 1933 and 1945, but during those years it was inspired by the challenge of depression and war—and by an exciting leader, Franklin D. Roosevelt. The trend to lethargy and self-protection, it seems to me, began with the end of the war and the death of FDR in 1945. Truman may be remembered fondly today, but in the late 1940s, he was not perceived as an exciting and challenging leader by federal workers. The loyalty program he established in 1947 was the first step toward making them value caution and keeping out of trouble. Then came the attack from Congress, from William Jenner and Joseph McCarthy, from the "who lost China" mob. The 1952 defeat of Adlai Stevenson, who as a candidate seemed to reawaken the idealism that had been dormant under Truman, was the final blow. The result was a near-total preoccupation with self-protection—a sense that the people at the top didn't share the civil servant's goals and didn't understand or care about his ideas unless the ideas were Red, which meant that he would be fired.

Any ardor that may have flared under John Kennedy or in the first years of Johnson's Great Society was snuffed out by Vietnam, so that by the late 1960s the bureaucrat was firmly set in the defensive ways that continue today—and have in fact been exacerbated by the Reagan administration's hostility toward government agencies outside the national security area. He doesn't want his performance to be evaluated because, although it might bring praise and more rapid advancement, it might also bring irrational inquiry into his political philosophy or the discovery that his performance has been less than superb. So wherever evaluation is attempted, he and his colleagues, allied in this cause as they are in no other, turn it into a joke. Early in the Carter administration, Barbara Blum, deputy director of the Environmental Protection Agency, tried to institute performance evaluations of that agency's employees. Back came the answer from the supervisors: 97 percent satisfactory. But Blum's supervisors were a bit more candid than most. The percentage of federal employees who got annual merit salary increases was more than 99 throughout the 1970s.

Because his performance is not going to be evaluated, at least not in any serious way, the civil servant knows that the only way he can lose his job is through a budget cut that affects his program. Therefore, since his survival depends on avoiding such cuts, it is in his interest to hide unhappy truths about his agency's performance from potential troublemakers, such as the Office of Management and Budget, which determines the president's budget recommendations to Congress; the General Accounting Office, which can audit the agency's performance and say it doesn't deserve what it is getting; and Congress itself, which has the final word on how much money the agency will get.

"No activity in a government agency is given as high a priority as securing and enlarging its budget," Leonard Reed, a writer and former bureaucrat, has said. "Bureaucrats almost invariably believe in the function their agency exists to perform, whether it is providing information to farmers or preserving the national forests. A new bureaucracy, the darling of the administration that establishes it, has a missionary zeal about its function. As a bureaucracy ages, it loses glamour and finds itself expending an increasing share of its energy on obtaining funds. . . .

"Gradually a hierarchy of administrative officers, executive officers, budget officers, congressional liaison officers, and public information officers grows up, almost the sole purpose of which is fund-wheedling. Since there is a certain logic to the proposition that without money an agency can't function, the bureaucrat . . . finds nothing wrong with spending more and more of his time and attention aiding the quest for more money, much of which is now needed to support the large money-raising apparatus that has grown up in the agency."

The effort to justify its budget often affects the way an agency does the job it was created to do. In 1974, the Internal Revenue Service began to single out middle-income taxpayers, who historically have the best record of paying their taxes honestly, for a disproportionately large share of audits. Why? Congress had granted the IRS more auditors on the condition that they be used to increase the number of audits performed during the 1974 fiscal year. The IRS chose to perform middle-income audits because they are quicker to do than, say, a full-dress study of Mobil Oil.

One of the most notorious results of the fear of budget cuts is the end-of-the-fiscal-year spending spree—remember, it is the fact of bureaucratic life the *Washington Post* reporter didn't know. It was a traditional rite of spring in Washington until a few years ago, when the end of the fiscal year was changed from June 30 to September 30. Now the season—harvest time—couldn't be more perfect, and the ritual remains the same. As the midnight hour approaches, each agency desperately tries to use up all its appropriated funds for that year so it won't appear to have been overbudgeted.

Several years ago Senator William Proxmire got hold of a memo sent by Admiral Elmo Zumwalt, then chief of naval operations, to his key subordinates, urging them to spend $400 million quickly before the fiscal year ended. When Proxmire held hearings on the memo and on the navy's efforts to get rid of the money by paying high claims to shipping contractors, Admiral Isaac Kidd, chief of the naval materiel command, explained: "We must . . . commit those funds within the prescribed period in order not to be put in a position of disadvantage later on by someone being able to say, 'Well, you asked for money, but you did not spend it, so we are going to take it away or cut your budget next year.'"

How did Admiral Kidd meet this crisis? "We have gone with teams of competent contract people from Washington to outlying field activities to look over their books with them . . . to see in what areas there is susceptibility to improved capability to commit funds." As every taxpayer knows, there is never a shortage of that susceptibility.

In its annual struggle to protect or enlarge its budget, a government agency has a decisive advantage over its natural enemy, the Office of Management and Budget, which appraises budget requests. The OMB might assign 6 or 7 employees to size up a cabinet-level department's budget, but the department will have 450 or 500 budget people whose job is to keep the OMB from finding out the truth. The agency may ask for four or five times what it needs for a program, confident that the OMB can't possibly track down all the overestimates, thus assuring that a more than satisfactory final figure will emerge.

Sometimes agencies do just the opposite: They *underestimate.* They do this, however, only for programs such as farm price supports and social security, in which the law requires fixed payments to qualified recipients. If the program runs short of money during the fiscal year, Congress must pass what is called a supplemental appropriation to take care of the shortfall or face a lot of angry constituents. The clever bureaucrat understands this: If, for instance, OMB says that a total of $9 billion is available for all his agency's programs, he might purposely underestimate the social security share, instead using some (or all) of it to fatten other programs, and go back to Congress later in the year to get the money to bail out social security.

If the OMB sees through all these tricks, there is yet another way of dealing with the problem—offering the OMB investigator a higher salary to come work for the agency. When this is done by a private government contractor, the impropriety is obvious. But it certainly has been done *within* the government.

There is one more budget-protecting play that often works even when the OMB recognizes it. I call it the Firemen First Principle. The basic idea is that, when faced with a budget cut, the bureaucrat translates it into bad news for members of Congress who are powerful enough to restore the amount eliminated. In other words, he chops where it will hurt constituents the most, not the least. At the local government level, this is most often done by threatening reductions in fire and police protection. At the federal level,

Amtrak showed it knew how to play the game: When threatened with a budget cut in 1975, it immediately announced that it would have to drop the following routes:

• San Francisco–Bakersfield, running through Stockton, the home town of John McFall, chairman of the House Appropriations Transportation Subcommittee;

• St. Louis–Laredo, running through Little Rock, Arkansas, the home of John McClellan, then chairman of the Senate Appropriations Committee; and

• Chicago–Seattle, running through the hometowns of Mike Mansfield, then Senate Majority Leader, and Warren Magnuson, then chairman of the Senate Commerce Committee.

And in a triumphant stroke that netted four birds with one roadbed, Amtrak threatened to cut a route from Norfolk to Chicago, running through the home states of Birch Bayh, then chairman of the Senate Appropriations Transportation Subcommittee, Vance Hartke, then chairman of the Senate Commerce Surface Transportation Subcommittee, Harley Staggers, then chairman of the House Commerce Committee, and Robert Byrd, then Senate Majority Whip.

The effectiveness of Amtrak's approach is suggested by a story that appeared in the *Charleston* [West Virginia] *Gazette* a few days after the announcement, under the headline "Continued Rail Service Byrd's Aim":

"Senator Robert C. Byrd, D-W VA, has announced that he intends to make an effort today to assure continued rail passenger service for West Virginia. Byrd, a member of the Senate Appropriations Committee, said he will 'Either introduce an amendment providing sufficient funds to continue the West Virginia route or try to get language adopted which would guarantee funding for the route for Amtrak.'"

In the Amtrak case, the bureaucrat's budget-cutting enemy was President Ford. Sometimes it is a frugal superior in his own department, but the trick still works. If, for example, a secretary of defense from South Carolina suggests eliminating useless bases to save money, the navy's bureaucrats will promptly respond with a list of expendable bases that is led by the Charleston Navy Yard.

The bureaucrat will almost always say that a budget cut is certain to result in the loss of jobs. The directly threatened employees will then write their congressman, who will almost certainly vote to restore the funds, because no one will have written him in support of the cut.

There is another aspect to the bureaucrat's concern about the loss of others' jobs. He knows he can't be a commander without troops to command. During the 1970s, the navy deprived its fleet of essential maintenance while continuing to spend billions on superfluous supercarriers and to trade small Polaris submarines for giant Tridents. The reason was that the more big ships with big crews there are, the more admirals are needed.

The same principle applies to the civil service, where rank is determined in part by the number of employees one supervises. Thus, a threat to reduce the number of a bureaucrat's employees is not merely a threat to his ego, but also to his position and income.

When menaced with a budget cut, the clever bureaucrat realizes that the public will support his valiant fight against the reduction only if essential services are endangered. Concentrated in the headquarters for bureaucracies of the New York City and District of Columbia school systems, for example, are some of the most prodigious do-nothing time-servers of the modern era. But no administrator ever threatens to fire them. They are the fat, and it would damage his cause even to admit their existence. Instead, he must concentrate on threatening a loss of muscle—the essential teachers. Similarly, the army, when faced with a budget cut, never points the finger at deskbound lieutenant colonels. The would-be victims are invariably combat troops. This practice is particularly unfortunate because in government, as in human beings, fat tends to concentrate at the middle levels, where planning analysts and deputy assistant administrators spend their days writing memoranda and attending meetings.

Memoranda and meetings are where the survival and make-believe principles merge. Bureaucrats write memoranda both because they appear to be busy when they are writing, and because the memos, once written, immediately become proof that they *were* busy. They attend meetings for the same purpose. Indeed, most bureaucrats make a big production of rushing off to meetings; meaningful action is seldom taken but the appearance of action is satisfied by the fact that the meeting was held. (Sometimes, attending a meeting is a prudent precaution against the possibility that bureaucratic enemies might use the occasion to cut your budget or lower your personnel ceiling.)

Naturally, the most favored meetings are those that involve travel. Here's an example from the travel records of the District of Columbia's departments of Human Resources and Education for the fiscal year ending in 1974: The Department of Human Resources spent $5,692 to send fifteen employees to San Francisco for the annual meeting of the American Public Health Association, and $4,882 for nine employees to attend an American Psychiatric Association convention in Honolulu. It also sent Raymond Standard, administrator of the Community Health and Hospitals Administration, to Aspen, Colorado, for thirty-nine days to attend a session on "Effective Strategies for Change." But his record was eclipsed by Dr. Jefferson McAlpine, administrator for the Mental Health Administration, who managed to stay on the road for fifty-one days. "If the same meeting that took place in Honolulu had been held in Baltimore," Dr. McAlpine told Thomas Love of the *Washington Star*, "there would have been no question on it." The trouble is that the meeting seldom is in Baltimore.

In fact, there were only three individual trips to Baltimore and a total of seventy-four trips to San Francisco, Honolulu, Montreal, San Diego, Disneyland, St. Thomas,

Quebec, Atlantic City, Williamsburg, Aspen, White Sulphur Springs, Miami, Tampa, and New Orleans. On group trips, Baltimore struck out completely; meanwhile, groups ranging in size from two to twenty-five managed to get to Squaw Valley, Ocean City, Denver, San Francisco, Virginia Beach, Atlantic City, Disneyland, Miami, New Orleans, San Juan, and Honolulu.

What would happen if we called the bureaucrat's bluff and cut his budget? Unless he thinks we know about the real fat, about those trips to Honolulu, he may—to preserve his credibility—actually fire essential employees. We could end up with a government run by planning analysts and friends of senators under whose rule trains would always run from San Francisco to Bakersfield via Stockton.

To cut the fat—*not* the muscle—out of the federal budget, there are four interrelated inflations to look at: inflated pay, inflated job descriptions, inflated grades, and inflated slots.

The last is easiest to explain: Suppose the country were being overrun by Albanian moths. The Department of Agriculture would ask the Civil Service Commission for authority to establish positions for, say, 200 Albanian Moth Control Officers. Because of the emergency, the authority would be granted and Agriculture would have 200 new slots. Suppose the Albanian moth were then brought under control. Would the slots be abolished? Certainly not. Now you understand slot inflation.

Then there is payroll inflation. Many federal employees are paid too much for the jobs they do. Payroll fat, as we have noted, is concentrated in the middle grades. In grades 13, 14, 15—where the responsibilities are usually quite modest but the pay is not—there are 200,000 employees earning from nearly \$35,000 to over \$55,000 a year. They receive, as we have pointed out, annual "merit" increases. In addition, they receive annual raises on the basis of maintaining "comparability" with private industry. But only the highest-paying 24 percent of private business is used in the comparison. The result is that, except in the very top jobs, the federal civil servant makes more than his counterpart in private business.

If government jobs paid as poorly as bureaucrats say they do, wouldn't there be a shortage of applicants and wouldn't bureaucrats be resigning in droves to accept the better-paying jobs in private industry? Neither, of course, is happening. In Washington, an average of 300,000 people a year seek government jobs. There are few positions to fill, however: Only 1 percent of those applicants ever find an administrative job—and I mean *any* administrative job.

The third kind of inflation—inflated grades—is connected to inflated pay. In 1977, the Civil Service Commission, not one of the bureaucracy's most severe critics, found that 150,000 workers were overgraded, meaning their salaries and ranks were higher than their level of work supposedly required. The commission estimated that these workers were being overpaid a total of about $780 million each year, because with their too-high grades come too-high salaries.

Behind the inflated grades lies the fourth kind of inflation: the inflated job description, the document by which the civil service system determines rank and salary. "Anybody who has ever worked in a government agency," writes Leonard Reed, "knows that job descriptions will endow a file clerk with responsibilities before which a graduate of the Harvard Business School quails." In even the smallest bureaucratic unit you will find at least one person skilled in writing job descriptions, who can turn typists into "word processors" and elevator operators into "vertical vehicle controllers." Those ten-dollar words produce thousand-dollar raises.

An embellished job description means, for the describee, not only more money, but also the chance of becoming a supervisor with a staff of his own, which means hiring new people who also have to be paid. The government's first supervisory grade is GS-13. Between 1965 and 1977, the number of GS-13s grew from 63,000 to 105,000.

Inflation in grade means inflation in space, as well. A

GS-6 gets 60 square feet of office space, while a GS-16 gets 300 square feet. It's an indication of the extent of grade inflation that in the ten years ending in 1980, while overall federal employment in the Washington area grew 25 percent, government-leased office space increased by 47 percent.

Reducing all this inflation will not be easy. Government employees can be fired by a process called RIF (reduction in force) if their agency's budget has been cut. But a RIF endangers the government's efficiency because employees are dismissed or retained on the basis of seniority, not ability. It was a demoralizing blow to the morale of the best civil servants when the Reagan administration used RIF to dismiss several thousand of them during 1981–82. But the administration had no choice of method of firing because over the years civil servants have managed to get regulations and court rulings that makes it next-to-impossible to fire them any other way. As one observed, "We're all like headless nails down here—once you get us in, you can't get us out." The Carter civil service reform bill made only a modest improvement in this situation, and the supervisor who wants to get rid of an incompetent employee continues to be faced with almost insurmountable obstacles.

One high-level bureaucrat claims that showing cause for the firing, as the law requires, can take up to 50 percent of his working time—and this for a period that may run from six to eighteen months. The executive may have to keep a diary of the employee's indiscretions (tardiness, mistakes, goofing off, flubbed assignments) before filing a complaint is possible. Protected by a maze of regulations and limitations and often defended by lawyers from government employees' unions, civil servants threatened with firing can drag out a hearing for months or even years. Eventually the boss may begin to feel *he* is on trial: His own reputation may begin to suffer as his colleagues and superiors accuse him of creating friction. At higher levels of government, as Leonard Reed notes, "harmony is valued well above function."

The result is a bureaucracy that is not only overstaffed but overstaffed with unproductive employees, employees who know that *the government's rate of discharge for inefficiency is only one-seventh of 1 percent.* Translated to a small business with ten employees, that's a rate of firing one person every seventy years. Carter's civil service reform made only a 10 percent improvement in that tiny rate of discharge.

What would happen if there were a dramatic reduction in the number of government employees? There is no conclusive evidence on this point, but scattered returns are suggestive. Have you noticed any difference in the service you get from Washington bureaucrats during the last two weeks of December? Probably not, but did you know that in recent years absenteeism among Washington federal employees during that time has run as high as 60 percent?

Also revealing is the discovery made by the District of Columbia government that the productivity of its trash collection crews increased when they were reduced from four men to three. And then there was the experience of the late Ellis O. Briggs when he was ambassador to Czechoslovakia. The Czechs became angry at us for one thing or another and ordered two-thirds of the American embassy staff sent home. Briggs found the stripped-down embassy the most efficient he'd ever had.

Probably the greatest obstacle to reform of the civil service is that most people think it is better to have a system based on merit hiring than one based on political patronage. But the fact is that getting a government job has only the most modest relation to merit. Veterans get five free points added to their civil service exam score; disabled veterans get an extra ten. For nonveterans the trick is to get their names requested from the Civil Service Commission by the agency filling the job, and the way to do that is to know someone inside the agency. People already in the system are the first to know about a job opening, and knowing both the applicant and the job, they can tailor the job description to fit the person they want to hire. So the civil service is a patronage ring based not on politics but on friendship. Insiders call it the "buddy system."

Even at the height of the Nixon scandals, Gordon Freedman, formerly of the House Manpower and Civil Service Subcommittee, contended that the real saboteurs of the civil service concept are the civil servants themselves. "Sure, some politicians get jobs for their friends," he said, "but you could put all the Nixon-referred people on the *Sequoia* [a small yacht used by Nixon] and it would still float. But if you put all the people involved in the buddy system on the carrier *Enterprise*, it would sink."

Another reason behind the opposition to political patronage is the national disdain toward politics and politicians. Philip Terzian expressed this attitude well when he wrote not long ago in *The New Republic* that "the pursuit of power is fundamentally a philistine occupation, and it is not likely that a genuine intellectual, mindful of history and human nature, would find the transient glory of public affairs worth the trouble."

We have an inexplicable regard for people who are "above" politics. James Forrestal, then Secretary of Defense, and one of countless possible examples, was praised by the *New York Times* for being above politics when he didn't support Truman in 1948, and both he and the *Times* were astonished when Truman proceeded to fire him. Forrestal was a Coriolanus. For him, being above politics really meant being above the mob. And this, I suspect, is the true attitude of most of the American elite.

It is widely assumed that a patronage system will result in a government run by unqualified people. Let's take a look at that assumption. Why do political employees *have* to be unqualified? A politically appointed typist could be required to type the same number of words per minute as the civil service typist. Remember that merit appointment and promotion are not the reality in the present civil service, it's only make believe. Friendship and a military background have a lot more to do with hiring and advancement.

Isn't it possible that government jobs might best be filled by politicians who are interested in putting together an administration that will do a good enough job to get them re-elected? The same principle applies to most of the decisions government employees make. Why shouldn't they be made on a partisan basis if the motive behind them is doing a good enough job to be re-elected?

This possibility has not been much explored in recent years. The same attitudes that produce civil service at the lower levels have led to the filling of higher level administrative jobs with people dedicated not to the success of the administration, but rather to their own progress up the meritocratic ladder from one prestigious post to another that is even more impressive. They are so intent on getting ahead that they usually serve in one job for only a couple of years before moving on to another.

This is particularly true at the assistant secretary level. Since World War II the average assistant secretary has spent twenty-one months at the job. This is unfortunate because, since the cabinet secretary has to spend most of his time worrying about the White House, the press, and the Congress and has little time for his own department, the assistant secretary level is where the government is actually run.

The main advantage of the career civil service is continuity. When I worked at the Peace Corps in the 1960s, there was a five-year limit on employment. The result was a steady, stimulating infusion of new blood and a much more adventurous group of employees than are attracted by the security of tenure. There was, however, a lack of continuity, and by the time I left, staff meetings had begun to seem like broken records. I heard problems discussed again and again as if they were brand new and the agency had no experience that would suggest their solutions.

There is another reason for not doing away with civil service tenure completely. Occasionally, unwise or corrupt political decisions may threaten institutions such as federal agencies, thus making it in the survival interest of the civil servants who work in them to blow the whistle on whatever wrongdoing is going on.

The role of the FBI and the CIA during the Watergate scandal shows how important the loyalty of the civil servant to his institution can be. When people in the White House wanted to contain the Watergate investigation, it was the civil servants who rebelled and leaked to the press. Indeed, the Watergate stories of the FBI and the CIA illustrate both the good and bad sides of the civil servant's institutional loyalty—both agencies showed an admirable, if only occasional and self-preserving, willingness to stand up to political authority gone wrong, coupled with a mindless and equally self-preserving dedication to covering up their own sins.

So instead of abolishing the civil service, perhaps 50 percent of federal jobs, as they open up through normal attrition, should be filled with appointees who can be fired at any time. (This doesn't mean, of course, that we should keep all the present jobs and just change people; it's clear that besides the problem of untouchable incompetents, there's a problem of jobs that are useless no matter who's doing them. It's a particularly thorny situation, because at least part of almost every job is useful—in some cases it may be only 10 percent, but because the Albanian moth is seldom completely eradicated, it's rarely nothing at all.)

Being able to fire people is also important, for two reasons: It permits the hiring of people we want and the jettisoning of those we don't want, and it makes it possible to attract to federal service the kind of risk-takers who are repelled by the political emasculation entailed by the Hatch Act's prohibition of partisan activity.

The key to democratic politics is accountability. If you don't deliver the goods, the voters can throw you out. Remember that when the Post Office was political it worked. We got our mail promptly. It was delivered twice a day, and packages arrived intact. Congressmen knew that if the postmasters they appointed didn't deliver the mail, the congressmen would be blamed by the voters. Now the congressmen can say that's out of their hands. And it *is* out of their hands. Congressmen have surrendered vast powers to independent federal agencies over which they and the president have little or no authority. Bureaucrats in such agencies feel beyond public control. Even when Congress or the president gives them an order, they find ways to subvert it.

The Library of Congress recently studied federal agencies' compliance with the Sunshine Act of 1976, which was supposed to open up government to the public. The study found that of a group of 1,003 government meetings listed in the Federal Register, 627 were either completely or partially closed to the public. One closed meeting was held by the Federal Reserve Board to consider the design of its furniture; it was closed on the grounds that "matters of a sensitive financial nature were being considered by the Board."

The military is a master of this kind of subversion. When the navy was ordered to conserve fuel during the energy crisis, it reported that it had reduced its ships' sailing time by 20 percent. What it actually did was redefine sailing time to exclude a ship's journey from port to the fleet at sea. In 1974 it was discovered that lunches served in the Pentagon's Army General Officers' Mess cost taxpayers an average $12.03 each. When exposed, the army announced reform: "Meal prices must be sufficient to cover operating expenses and food costs." Did the generals give up their $12.03? Not on your life. They redefined expenses to exclude such items as stoves, utilities, and waiters' and cooks' salaries.

What is this if it is not make believe? Laws are passed,

orders are given, compliance seems to occur, but nothing really changes. Bureaucrats don't like real change, only the appearance of change. That is why they are so fond of reorganization. Reorganization gives them something to do—redrawing charts, knocking down office walls—but nothing outside the agency, such as poverty or hunger or disease, is affected in the slightest. What does happen is that new jobs are created, almost always with higher grade classifications, which of course means higher salaries for the reorganizers.

The reason why bureaucrats like internal reorganization better than external action is easy to understand. Suppose you work in an antipoverty agency and you do your job so well that poverty is eradicated. Or suppose you work in the Department of Energy and the energy problem disappears. What will happen to you? The bureaucrat can figure that out. If he takes real action, if he's truly effective, he'll be out of work, he won't survive. If, on the other hand, his action is make believe, poverty will not disappear, the energy problem will not be solved, and his job will be safe—he will survive. Now you understand the fundamental Washington equation:

Make believe = Survival

FOREIGN SERVICE OFFICER
UNITED STATES of AMERICA
SPEAK ENGLISH ONLY

4 The Foreign Services

"Burns admired professional diplomats—men who were cool, collected, in control. . . . He saw himself as one of a dozen men in . . . an anteroom in a foreign chancellery (Belgrade? Helsinki?) conducting secret negotiations, ADC to a giant, Bohlen or Kennan, taking on the Russians by sheer force of logic and remorseless dialectic, arguing them back. . . . Forcing an agreement, and then a laconic cable to the Department. *Negotiations concluded.*"

Thus Ward Just captures in his short story, "Burns," the life foreign service officers dream about but seldom lead.

It's not that their lives lack adventure. "A friend of ours," wrote Bill Keller and Ann Cooper in *The Washington Monthly,* "is a cultural officer in the foreign service. We see him occasionally on his holidays, and he always has terrific stories that fire up our sense of adventure—stories of crossing the Sahara with Bedouins, riding the White Nile steamer from Juba to Khartoum, dodging mortar fire in Lebanon."

What is missing is significant work. Those secret negotiations seldom take place. As Keller and Cooper noted, their friend "never talks about the *work* he does, but we always assumed that this was just a diplomat's discretion in the presence of journalists. Once, however, at a Thanksgiving reunion of old friends, we asked him for a specific example of what it is, exactly, that a cultural officer does. There was a long, hesitant pause, and then he came up with one: He had once arranged the visit of an American bluegrass band to Kuwait."

In comparing the image of life in the foreign service with the reality at the U.S. embassy in Morocco, Keller and Cooper found:

"The international intrigue, the classified cables, the ringside seat at the unfolding of great events—this is the glamorous image of diplomacy, an image carefully nurtured by generations of State Department officials. Close up, the picture is very different; our experience suggests that on an overseas tour of duty the typical diplomat lives a life largely consumed by make-work, devoid of genuine responsibility, and contributing little to the advancement of America's interests abroad."

In other words, the typical diplomat is much like his fellow government employee in Washington. His main activity is make believe, and, as we shall see, his main purpose is survival.

The State Department is one of three sizable civilian—we will deal with the military in the next chapter—bureaucracies involved in foreign affairs. Another—the National Security Agency—breaks other countries' codes and is supersecret. The third is the Central Intelligence Agency, from which a good deal of secrecy has been stripped away. There are also smaller organizations, such as the Agency for International Development, the International Communications Agency (formerly the United States Information Agency), and the Peace Corps. This is not to mention various minor contingents from the FBI, the Drug Enforcement Administration, the departments of Agriculture and Commerce, and a few other agencies that manage to get a foot in our embassies' doors.

While the bureaucratic characteristics of these organizations are similar to those we have found in domestic agencies, some are unique to the foreign affairs area—or can be found there in either larger or more dangerous doses.

One such characteristic is clientism. This is the tendency of our representatives to inflate the importance to the United States of the country in which they are serving. A dispassionate observer might conclude that cabinet changes in Burma or Paraguay are of little importance to us. But you can be sure that the members of the U.S. country team, which is what the group of senior representatives of the various American agencies in the host country is called, are dramatizing every development with cables to Washington marked "Top Secret" and "Eyes Only." Their reasoning is that if Washington thinks the host country is important, it will think *they* are important, with a resulting rise in their prospects for the various promotions, raises, and honors they know in their hearts they so richly deserve.

Clientism can take another form: identification with the host country or with those currently in power in the host country. Lyndon Johnson used to complain that his ambassador to India, Chester Bowles, was really another ambassador *for* India, seemingly more interested in advocating India's cause than America's.

This kind of clientism can have a darker side: In 1969 Thomas Melady was our ambassador to Burundi, a small Central African country with a history of tribal bloodshed between the Tutsi, the dominant minority, and the Hutu, who made up 85 percent of the country's population but had been denied political and economic power. The Tutsi feared that the United States would take sides with the suppressed Hutu, but Melady set about overcoming his clients' concerns. "He told them every chance he got," remembered one American official, "that the United States was absolutely impartial . . . that their relations were their own affair, and he apparently got through to them." U.S.–Burundi relations were never better. Then in May 1973 the Tutsi murdered a quarter-million Hutu.

Melady worried about reporting the genocide to Washington, afraid that the State Department would somehow "over-react" and destroy his carefully nurtured relationship. He arranged a letter to the Burundi government from several members of the diplomatic corps. "It was a low-key thing," one embassy staff member recalls, "saying we were concerned with their difficulties." Another remembered it as "tactful . . . it got no real response."

There is little doubt about Melady's motive. "He wouldn't sacrifice the relations he's built up," reported the source to Roger Morris, who repeated it in an article in *The Washington Monthly* several years ago. Ronald Reagan has since rewarded Melady by making him assistant secretary of education.

Another truth common to all the civilian foreign affairs agencies is that no one wants to get stuck in out-of-the-way places such as Ouagadougou for more than two or three years. This is the main reason for the frequent shifts of personnel that are characteristic of each of these agencies. If you've been in Ouagadougou (it's the capital of Upper Volta), they try to give you Paris or Rome next, or at least a pleasant African city like Nairobi.

One result of these frequent transfers is that too few foreign service officers have the time or the motivation to learn the local language and culture. This problem was pointed out dramatically more than twenty years ago in a book called *The Ugly American.* Yet, as late as 1978, only one in ten of the foreign service officers stationed in Iran was even minimally competent in Farsi, Iran's principal language. At the time of the Sadat assassination in 1981, only 59 of 800 American officials in Cairo spoke Arabic.

Frequent transfers are also bad for institutional memory. Since there is a new staff every few years, no one can remember what happened more than a few years ago. In the early 1970s, William Paddock, who was writing a book about the effectiveness of American foreign assistance programs, returned to Guatemala to visit the AID mission where he had worked nearly twenty years before. He talked to the mission's director, William Hinton:

Paddock: I understand Barcenas includes the forestry school

the U.S. government helped establish ten years ago and later helped merge with the agricultural school there.
Hinton: I don't know anything about that. You must remember that I have only been here fifteen months. There is a lot about previous programs I don't know.
Paddock: Is any money going into the experiment station at Barcenas?
Hinton: What experiment station? There is no experiment station there in the sense any of us would think of one. It's a work farm for the Barcenas students. . . .
Paddock: I don't mean the school's farm, I mean the experiment station. When I worked here in the 1950s this and the station at Chocola formed a major U.S. government effort. . . .
Hinton: I know nothing about it. I'm still learning.

Another basic truth about the foreign affairs bureaucracy was uncovered by Paddock when he asked Covey Oliver, then assistant secretary of State for Inter-American Affairs, to select a successful project for Paddock to visit. Oliver recommended Los Brillantes in Guatemala, where he said AID was succeeding in helping Guatemalans break the bonds of one-crop (coffee) dependency by supplying seedlings, advice, and loans to encourage the planting of rubber, citrus, and other crops. When Paddock reached Los Brillantes, "the place seemed dead." There was only one AID employee there. The loan money had run out. None of the farmers were planting rubber or citrus.

Why had Oliver thought the project was a success? The foreign affairs bureaucracy, like bureaucracies in general, tends to gild the lily as information travels from the field to the home office. A project that is a disaster will be presented to top Washington officials as an outstanding success. This tendency toward make believe may be greater in the foreign service because of the physical distance that separates the bureaucrat in Washington from the truth in the field.

The reason for lily gilding is that subordinates want to please their superiors, which is also why subordinates don't speak up when their opinions might not be popular with the people at the top. William A. Bell, a former foreign service officer tells these stories of the 1960s:

• In 1966, when the commitment of American ground forces in Vietnam took its greatest leap forward, criticism of U.S. policy became widespread among foreign service officers, or at least among those stationed in Washington. A number of young officers, some of whom had been privately expressing their misgivings, were called together for a briefing before setting out on campus recruiting trips. One of them asked the recruitment director what they should say to students who were interested in the foreign service but had qualms about the American role in Vietnam. The answer—in no uncertain terms—was that there was no place in the foreign service for persons who do not support the war. No one spoke.

• In 1965, at the beginning of the rebellion in the Dominican Republic, U.S. ambassador W. Tapley Bennett declined a request by the opposing parties to mediate the rapidly growing dispute at a time when moderate leftists were still in control of the "constitutionalist" forces. Bennett's predecessor, John Bartlow Martin, states in his book, *Overtaken by Events,* that Bennett, having missed this chance at conciliation, probably had little choice but to bring in the Marines.

The book fails to relate, however, a scene in which Bennett summoned a large portion of his staff and told them that he was planning to call for help. After briefly describing the situation as he saw it, Bennett made it clear that U.S. military forces, if summoned, would be ordered to thwart the attempted revolution, not just "protect U.S. lives and property." He then asked his staff if there were any alternative views or proposals. No one spoke.

• When John Bowling, a stimulating lecturer at the Foreign Service Institute, suggested that flag desecrators were philosophically identical to the bomb-throwing anarchists of previous decades, and that draft resisters were unmanly and cowardly, not one of the foreign service officers in his audience challenged the statement, despite Bowling's invitation to do so. No one spoke. After several moments of silence, Bowling himself finally felt compelled to express the other side of both positions.

Another illustration of how employees in the field are influenced by what their bosses want to hear comes from Patrick J. McGarvey, who used to work for the Defense Intelligence Agency. From 1964 through 1966, when the generals wanted excuses for building up American troop strength in Vietnam, the DIA flooded Washington with reports of growing communist strength. Then in 1967, the generals decided they wanted to show success. And Washington was again flooded with cables, this time describing enemy body counts and pacified villages. If your boss is a John Foster Dulles who does not like Sukarno or a Henry Kissinger who does not like Allende, you tend to find intelligence that says Sukarno and Allende are bad guys.

And if the boss wants his agency to expand, information that justifies expansion is what he'll get. When I worked for the Peace Corps during the early 1960s, Sargent Shriver was eager to build up the agency as rapidly as possible. People called Programmers were sent out to foreign countries to solicit invitations for volunteers. Since many host country officials were anxious to please the brother-in-law of the new American president, they were liberal in issuing invitations. And since each Programmer was anxious to return to Washington "with a program in his pocket," as the saying went in those days, the invitation was not always subjected to careful scrutiny. The result was that hundreds of volunteers were dispatched to fictitious jobs or to places where they weren't really needed.

The tendency to not tell the boss what he doesn't want to hear is of course true of organizations generally. In the foreign affairs area, however, its consequences can be a catastrophic loss of life. It is doubtful that the Bay of Pigs would have occurred had someone had the courage to tell

President Kennedy that cancellation of the air strike, which was designed to wipe out Castro's air force, meant that the invasion could not succeed. But no one spoke.

Joseph Burkholder Smith, a retired CIA official, says, "Policymakers will ignore intelligence that shows they have taken the wrong course of action, and the CIA station will oblige this inclination by providing intelligence that shows the policymakers they were right." Thus, when South Vietnam was collapsing, in March and April of 1975, Henry Kissinger and Gerald Ford didn't want to face what was happening, so Graham Martin, the U.S. ambassador, and Tom Polgar, the CIA station chief, kept feeding them overly optimistic reports. The result was that we did not plan an orderly evacuation and left behind tens of thousands of Vietnamese who had been led to rely upon us.

The man who told the truth about this episode was one of Polgar's subordinates, Frank Snepp, in a book called *Decent Interval.* The CIA sued Snepp. It promoted Polgar, leaving his colleagues an indelible lesson in how to get ahead. Getting ahead is a factor in another characteristic of the foreign affairs bureaucracies—their tendency to look inward and upward rather than outward to the people of the country they are serving. Their survival network does not consist of campesinos sweating in Panamanian jungles or the starving poor of Calcutta; it is made up of other foreign service officers who will sign their fitness reports and sit on their promotion boards and of other Washington officials who can influence them. Their primary concern is with Washington, where the higher officials are. The drama of the day comes from reading cables from Washington and preparing cables to Washington—changing one word can cause hours of debate. The only thing more important is Washington in the flesh in the form of a high-ranking visitor. Then, the embassy throbs with vitality. The young aide sitting in one of the jump-seats in the ambassador's limousine and listening to the ambassador exchange anecdotes with the visiting undersecretary waits expectantly for the moment when he can interject the incisive comment that will make him be remembered as "that promising young officer in Dakar."

Second in importance to the Washington connection are relationships within the embassy itself. The process that John Kenneth Galbraith, who served as ambassador to India, described as operating in the State Department also prevails in embassies:

"When I went back [to the State Department] this time one of my assistant secretary friends attended the Secretary's staff meeting from 9:15 until 10:00 A.M. Then he had a meeting with the undersecretary on operations until 10:30. Then he took until 11:30 to inform his staff of what went on at the earlier meeting. Whereupon they adjourned to pass on the news to their staffs."

Day after day is spent in the make believe of meetings because there is so little else to do. Almost all of our embassies are overstaffed, as is the State Department itself. We used to have an ambassador to Chile who reported directly to the secretary of State. But soon a Chile desk officer was interposed in the hierarchy, followed by an assistant secretary for Latin America (with deputy), and then a regional director for the West Coast of Latin America, with his deputy. So now there are at least six layers of titled officials between the secretary and the ambassador. And of course each embassy has been stratified so that the ambassador presides over an elaborate chain of command, at the bottom of which is the junior foreign service officer who actually writes the first draft of the report.

All that's really needed is someone to write the report and the secretary to read it. But the bureaucratic tendency to expand has generated a stepladder of well-paid officials. If they did nothing at all, that would be bad enough. But, of course, they have to add something to every report to justify their titles, and in the process reports become so boring and devoid of meaning that the secretary stops reading them.

Third in importance in the life of the embassy is the relationship between the various elements of the American community. Interagency rivalries, such as those that have from time to time existed between State and the CIA and between the Peace Corps and AID, consume tremendous amounts of time. Sometimes they may even have cost lives. For example, for years the foreign service insisted upon distinguishing itself from the CIA in the *Biographic Register*, a kind of studbook published by the State Department. This was done in such a way that a hostile agent could easily identify the CIA's people. The practice was stopped only after Richard Welch, a CIA station chief in Athens, was murdered in 1975. Fortunately, these interagency rivalries are usually more benign. In the mid-1960s in the Cameroons, the debate at one embassy meeting was on how to divide the shipment of Skippy peanut butter that had just arrived at the embassy PX.

The best example I know of the psychology of those who serve in U.S. missions overseas comes from this letter written to a friend of mine by an employee of AID in Saigon (names have been changed):

"Shortly after my return, we finally moved to USAID II! Delays of a couple of months in the schedule, but made it by mid-August. Much nicer quarters and location. I ended up with a nice little office, carpeted, etc. next to ADCCA with windows looking over the yard and entrance and shaded by the trees. . . . While I handled the overall coordination, when it came to space allocation, I just tried to see that each division received about the same amount of square footage on a per capita basis, and let each Assistant Director decide how he would use it. Passing the buck nicely. I remember how much time and worry you had spent on that and so was looking for an easier way. Brother Art Mason came out with an office the same size as mine (about 120 sq. feet), but located in the middle of CDI (cap. dev. and industry as it was named after you left). This made him very unhappy and he tried to get located on the first floor close to ADCCA instead of the ground floor with the industry boys.

"When I returned I found the office in the midst of preparing IRRs on about nine capital projects for FY 75 presentation. . . . After that was completed we prepared four CAPs at the same time (rural credit $60 million, industrial credit $40 million, Saigon Export Processing Zone $5 million and low lift pump irrigation $6 million). By the time that was completed, into preparing the documents for FY 76 program (now in place of IRR, two separate papers are needed to get the item into the Congressional Presentation), begun. Without knowing what Congress would do to us in FY 75, we requested 14 projects totaling about $260 million! That exercise was completed by January 30. . . .

"The agency is going through a major RIF which is being mishandled in every way possible. . . . The names finally came out Monday. Neil Anders and Eddie Jackson are the only two in CDI. Neil could retire and was expecting it; Eddie can't retire and his was unexpected by everybody. . . . Joe Falcone survived termination of his appointment by postponing actions for many months while filing about five grievances (several were valid, for the personnel people mishandled every aspect of his case), and is being transferred to Haiti as C.D. officer there. Ben Wyle is going to Abidjan. Jim Fisher is expecting and hoping to go to ROCAP (in Guatemala) for a tour that will set him up for retirement."

This letter was written in March 1975 as South Vietnam was falling apart. There is no mention of what was happening to the country in which the writer was working. Instead, as the South Vietnamese Army was collapsing in the Central Highlands and the end for all of South Vietnam was only six weeks away, he writes only about budget, personnel, and the square footage, floor covering, and view of his office.

This is typical of the way far too many foreign service people think. Concerns internal to the mission are dominant. And of those, personnel policy is the most important. Here the AID employee is worried about reductions in staff, until Reagan an uncommon event in government. More often the concern is about promotion. In the foreign service, for example, the shortage of significant work, combined with the reluctance of foreign service officers to quit, means that people must wait a long time before they can move from one of the make-work jobs into one of the few with challenging duties.

John Kennedy's staff used to tell him that if he had gone into the foreign service, he would, at the same age he became president, still have been an FS-2 or FS-3, ten years away from an ambassadorship. Richard Neustadt says there are fifteen years between the time a foreign service officer is ready to assume responsibility and the time he gets it. Usually this gap covers the period from age 35 or 40 to age 50 or 55.

Why do they stay? One motive is the prestige of the foreign service, the reluctance to give up that institutional badge; this is the same reason that so few Rhodes scholars come home early. In fact, as James Fallows has pointed out, the foreign service is a kind of life-long Rhodes scholarship, a badge of status more important for the identity it gives the wearer than for what it requires him to accomplish. The badges confer a special sense of class, almost a touch of aristocracy. The little secret that no one lets out is that what one *does* after putting on the badge is not all that exciting.

Most young entrants say they are joining the service because they want to influence foreign policy. To show what happens later, John Harr's book, *The Professional Diplomat,* quotes one official: "The trouble with most FSOs is that they are too concerned with *being* something or *becoming* something—*being* a deputy chief of mission or *becoming* an ambassador—and not enough with *doing* anything."

A 1967 White House memorandum discussed the years foreign service officers spend waiting for responsibility: "[It] is a long wait. It is a period during which most officers will make at least a small blot on their copy books. The trauma of a bad (or mediocre) efficiency report is ordinarily enough to impress upon the recipient the value of caution and patience. Mid-career officers come to appreciate that the use of the word 'abrasive' once in an FSO's files can be enough to counteract repeated appearances of words like 'creative' and 'resourceful' . . . so the art of becoming un-abrasive becomes part of an FSO's stock in trade." When William Macomber, then deputy undersecretary, testified before Congress in 1971, Senator Claiborne Pell asked him what an officer should do if he were asked to carry out a policy he thought wrong. Macomber replied: "If he is the kind of person that has a pretty low boiling point on these matters, if he really feels untrue to himself to compromise in that way, then I think he is in the wrong business. Then I think he ought to be in politics, speaking out, or be a teacher or writer. Thus I think you have to accept certain inhibitions if you accept a career in the foreign service. On the other hand, there is a marvelous reward if you can stay in the foreign service and live with the kind of inhibition I described. Because then you are guaranteed a ringside seat in this terrific effort to make peace in the world."

Unfortunately, a ringside seat, watching rather than participating in crucial events, is all that most foreign service officers ever get. Usually they must settle for carrying the secretary's briefcase or looking over his shoulder during the great moments in diplomacy. But for the handful who finally do achieve Burns's dream and send that "negotiations concluded" cable, the satisfaction is immense.

Another reason foreign service officers like being foreign service officers is that they become closely associated with Harrimans and Lodges, with the elite group that has filled most of the cabinet and sub-cabinet jobs in the foreign affairs area for most of our history. The members of this elite tend to have what William Colby has called "impeccable social and establishment credentials." They write for *Foreign Policy* and *Foreign Affairs.* They move back and forth between government, law and investment firms, and the foundations. "Theirs is a life," Roger Morris has written, "lived mainly in carpeted offices and quiet board rooms, well insulated from the rest of the country."

When there are such people in charge of foreign policy and foreign service officers working for them, there isn't likely to be any dramatic break with the past. Indeed, it is much more likely that past errors will be defended or protected from exposure.

The CIA used "national security" to justify what was nothing more than bureaucratic self-protection. For example, it tried to suppress the fact that the following sentence was part of a report written one week before Syria and Egypt invaded Israel in 1973: "The movement of Syrian troops and Egyptian military readiness are considered to be coincidental and not designed to lead to major hostilities."

There have been plenty of similar errors. Perhaps the most dramatic during the late 1970s was the failure of the State Department and the CIA to anticipate the revolt against the Shah in Iran. Here the significant factor may have been the bureaucrats' tendency to rank personal pleasure above the public interest. In Iran, this involved staying in the capital city, where a pleasant social life could be enjoyed and where, until the last moment, one either did not hear of the discontent stirring in the provinces or did not realize the potential magnitude of the uprising.

Indeed, it might be said that the true guiding principle of the foreign service and the CIA, also followed religiously by most American journalists overseas, is Never Leave the City Where the Good Bars Are. This principle has relieved the monotony of our lives by contributing greatly to the frequent occurrence of the unexpected in the Third World, from Jonestown to Vietnam.

Where you *can* find our foreign services are places like London, where our embassy houses a staff of almost 700 people. Fifty-two of them work for the International Communications Agency. Their function is propaganda, promoting understanding of and sympathy for American policies, valiantly restraining the virulent anti-Americanism of the English people.

If our foreign services waste their time on this kind of activity, why have them? An official British study group has recommended abolishing Her Majesty's diplomatic service. We can't go that far—although it is tempting, isn't it—because we do want the kind of Peace Corpsmen who actually help other people and the kind of foreign service officers and CIA agents who give us information we really need about other countries. But we could end either State Department or CIA representation in countries from which we do not need duplicate reports and abolish both in countries of little or no importance to us. Where minimal facilities are needed to take care of lost tourists, we could make agreements with friendly countries so that, for example, we would represent Canada in Chad and Canada would represent us in Mauretania.

If even this seems too radical a departure, remember Ambassador Ellis Briggs's experience in Czechoslovakia, where he found his embassy performed more efficiently after most of the staff had been sent home. Bill Keller and Ann Cooper learned the same lesson when they went on from Morocco to visit the much smaller American diplomatic community in Mali. A young AID economist who was serving in Mali but had also worked in Morocco told them: "Here I write a cable, and if it's correct and reasonable, it goes. It's just that people here have more things to do to occupy themselves than pick cables to pieces. In Rabat, everything was make-work."

Newsweek, in its issue of October 29, 1979, reported that "six U.S. government officials abroad have been killed and thirty-four more involved in terrorist attacks or kidnappings so far this year." The very next week came the seizure of the hostages in Iran. Should we ask people to risk their lives for the sake of make-work?

USS CAY

5 THE MILITARY

Vietnam was a scarring experience for the nation, but it was crippling for the military. The enlisted man's trust in and respect for his officers—that crucial connection between the leader and the led—was badly frayed.

Even before Vietnam, there were traditional—but grating—differences between officers and enlisted men. An officer, for example, who is retired before completing twenty years of service can collect up to $30,000 in separation pay. An enlisted man gets nothing. Visit almost any military base and see how much better officers' housing is compared to the enlisted men's. And note the often luxurious officers' clubs compared to the relatively simple social facilities of the enlisted men. When things go wrong, the officers usually do the investigating and the enlisted men usually take the blame. Of the first 126 soldiers relieved from duty during the 1979 army recruiting scandal, only 3 were officers.

The differences in the way officers and enlisted men live are illustrated by this passage from Stuart Loory's book, *Defeated: Inside America's Military Machine*: "In the winter of 1971–72, Rear Admiral James E. Ferris commanded an aircraft carrier task force on Yankee Station in the South China Sea washing the coast of North Vietnam. . . . Ferris's quarters, known as 'admiral's country' aboard ship, were . . . like a New York Central Park South luxury apartment. . . .

"A huge green plant grew against one wall. . . . The planter rested on thick beige carpeting that helped deaden the bone-shaking noise of aircraft launching and recovery operations. There was not a stray bit of dust on the glass-topped coffee table nor an ash in any of the heavy glass ashtrays. The thought was inescapable that orderlies were ever present, waiting for the tiniest bit of refuse to accumulate, sweeping it up as soon as the admiral left the room. On one wall, a walnut credenza held a stereo tape player. The admiral's desk was set against the opposite wall.

"The sofa was covered in a nubby white fabric. It was functional but soft, enveloping a visitor in instant comfort and security. Just outside the doorway to the study, a Marine aide stood attentively, if not always at attention, waiting to carry out the admiral's every wish. In the small galley, two Filipino messmen were preparing a chicken dinner for the evening.

"The admiral's dining room . . . could seat ten comfortably around an oval table covered with starched white linen. The silverware was heavy and glistening. The meals were served with painstaking etiquette by white-coated attendants. . . .

"Belowdecks, the crew was jammed together, 150 men to each open, windowless, poorly lighted, ill-ventilated bay. They lived one atop the other, three bunks high, with no privacy and little storage space, with the constant noise of the ship's operations jarring them. They took their meals in windowless, low-ceilinged mess spaces that doubled as warehouses for the bombs and rockets the airplanes would use."

All the resentments that such experiences produced were exacerbated by the contrast between the enlisted man's twelve months on the firing line in Vietnam and the five- or six-month combat tours for officers. Platoon leaders and company commanders were "rotated" in and out of units before they had a chance to get to know their men or to master their jobs. Only 3 American generals were killed by enemy fire in Vietnam. By contrast, 223 German generals—a *third* of the total number—were killed during World War II. The casualty rate among German generals was higher than among enlisted men. In Vietnam, the reverse was true, and by a grotesquely wide margin.

"The cohesion of a combat unit," wrote Richard L. Gabriel and Paul L. Savage in their book, *Crisis in Command,* "is to a large extent a function of the degree to which combat troops perceive that their officers are willing to fight and die with them." But in Vietnam, the higher one's rank, the lower the probability of death on the battlefield. The result was that the morale of the enlisted men was terrible. They took drugs, refused to obey orders, and even rolled grenades into their officers' tents, a custom known as "fragging."

Vietnam, because of its unpopularity, also had an unfortunate effect on the social composition of the military. The proportion of college graduates in the services dropped 75 percent between World War II and Vietnam. Now enlisted men come from the poor and officers from the lower to middle range of the middle class. The result is a military that does not represent the variety of talents and points of view of the country as a whole, a military that has almost no connection with a significant part of the population—the very part of the population, in fact, that tends to produce the leaders of the rest of our institutions.

One evil consequence of this became dramatic during Vietnam: The lower classes, scorned by their betters, who had the wit and resources to avoid the draft, did almost all of the dying. Another is the general ignorance of the educated elite about military affairs, because its members have not served in the military or even had friends who have done so. Thus, when the question arises as to what kinds of planes, tanks, and missiles the country needs or what the role of the army or the air force should be, intellectuals tend to adopt, through lack of real knowledge, clichéd positions that are usually stupidly antimilitary but are sometimes equally stupidly pro.

Many of the villains behind the deterioration of the military are not easy to identify, rooted as they are in the complex sociology of modern America. One, however—the reason for those five- or six-month tours for officers—is very clear indeed: The officers were getting their "tickets punched," the term that describes what an officer has to do to get ahead in the military. The most essential ticket punch of all is combat command. There was tremendous competition for troop commands in Vietnam, from platoon to company to battalion to regiment to division. To give more officers a chance to serve in a combat job, tours were limited to a few months each.

The military has refined "ticket punching," its version of civilian society's meritocratic scramble, to a science. You are more likely to get ahead as an officer if you went to West Point or Annapolis or the Air Force Academy. West Point graduates constitute 11 percent of all army officers, but they make up 46 percent of the generals.

In the army, here are the other ways of getting your ticket punched (there are comparable methods in the air force and navy):

• You should attend the Command and General Staff College at Fort Leavenworth, Kansas. This is the most prestigious of all the army's staff colleges. You are selected for it by a special board when you are a captain or major. Class rank here is even more important than at West Point because here you're studying bread-and-butter issues in military strategy and tactics and your record will follow you throughout your career.

• You should work at the Pentagon. There you will have the opportunity to meet powerful figures from the Congress, the Defense Department, and other parts of the executive branch. This is where you attract the patrons and sponsors who will make sure that you get choice assignments. Alexander Haig was the modern master of this art.

• You should get a graduate degree at a civilian university. This "requirement" came into existence during the days of Robert McNamara, when being an intellectual whiz kid was the thing. Almost all generals have one or more such degrees today.

• You should serve in a division with a good reputation, such as the 82nd or 101st Airborne. Membership in such units suggests that you're likely to be fit, competent, and brave.

• You should attend the National War College. Located on the grounds of Fort Leslie McNair in the southwest part of Washington, D.C., this school attracts high-level civilian officials as well as lieutenant colonels from the army and air force and commanders from the navy. The curriculum consists not of military tactics but of larger political and economic issues that will prepare the students for high-level participation in global strategy and national security policy.

All of these experiences contribute to your survival network. Indeed, the construction of that network is what ticket punching is all about.

Here is the story of how one general built his network, as told by the *Washington Star*'s John Fialka:

"General Bernard W. Rogers, named by President Carter to succeed General Alexander M. Haig, Jr. as supreme commander of NATO, . . . graduated from West Point in 1943, . . . started out as a platoon leader in the 70th Infantry Division. . . . General Maxwell D. Taylor, . . . looking around for a young officer to assist him as commandant of West Point, . . . asked who [was] the brightest and most promising officer, and 'four people told me Bernie Rogers,' recalls Taylor. In 1947 Rogers was awarded a Rhodes scholarship and went to Oxford University in England, where for three years he studied philosophy, politics, and economics. . . . He saw his first combat in Korea in 1952, when he won a silver star as the commander of a 2nd Infantry Division battalion. From the late 1950s through the mid-1960s, Rogers held a number of key staff positions in the Pentagon, culminating in 1963 when he was appointed executive officer to Taylor, then chairman of the joint chiefs.

"From November 1966 to August 1967, he [was an] assistant division commander of the 1st Infantry Division in Vietnam. After that . . . commandant of West Point. In 1971 Rogers . . . helped insure his ascendency. He became chief of legislative liaison in the office of the secretary of the army and gained the respect of a number of powers on the House and Senate Armed Services committees."

There are both good and bad aspects to the present system of ticket punching. It's good for officers to know that command of combat troops is important to their careers. It means that, as they rise in responsibility, their decisions will reflect firsthand knowledge of what life is like for combat units. And because they have been exposed to this true performance test they are more likely to be skilled than the average civil servant of comparable rank, who can escape such tests throughout his career. But it is not good for this experience to be sought, as it was in Vietnam, at the cost of those five- or six-month tours that were so destructive to the morale of the troops. Nor is it pleasant that the system makes young military officers pray for a war in order to get ahead.

And, as we have already seen with the foreign service, the frequent change of assignment required by the ticket-punching system produces low institutional memory and repeated discoveries of the wheel at each post. There is an accompanying lack of interest in long-range improvement of any unit. The officer cares about how it looks today so that he will look good, but not about how it will look tomorrow, because he won't be there tomorrow. And because he won't be there tomorrow he does not develop the loyalty to and pride in his unit that throughout history produced outstanding battalions and regiments.

A farcical aspect of the ticket-punching system is the Officer Efficiency Reports (OER), a series of performance ratings that each officer gets every six months as he progresses from second lieutenant to major general. The OERs, like the dollar, grades in college, and annual merit ratings that produce raises for almost all civil servants, are badly inflated. "Commanding officers know," Nicholas Lemann has written, "that good scores make for happy subordinates and that bad scores, rather than serving notice of temporary failings, can wreck careers. So the maximum score of 200 is common and 185 is a disaster. If you average below 195, it's thought impossible to make major.

"To be sure, there are subtle ways a rater can damn an officer without having to give him a low rating. For instance, a rater can say in his typed comments on the officer that he did 'a superior job' and leave his career in ruins. Elsewhere on the report 'superior' appears as the second-best adjective in a hierarchy of praise that's led by 'outstanding,' so to call an officer superior is to call him second-rate. Also, promotion boards go through OERs so fast that any reservation-expressing word like 'but' or 'despite' or 'although' is thought to stand out as a red flag of disapproval regardless of its context.

"Because no officer who wants to advance himself can afford a bad report, no officer can afford to cross his boss."

As I have pointed out, from the moment the top officials at the Pentagon instituted the body count as the measure of success in Vietnam, commanders began to pressure their subordinates to produce bodies. And because the subordinates wanted to please their superiors, they began to produce a lot of bodies, which meant shooting a lot of people, including civilians. At My Lai, somewhere between 175 and 400 Vietnamese noncombatants were slaughtered by soldiers of the American division under the leadership of General Bernard Kosters, who, along with his key subordinates, immediately instituted a cover-up that took investigators under Lieutenant General William R. Peers several years to expose. Peers's final report identified thirty officers who were guilty of the cover-up. Of these, only four were actually put on trial by the army, and three of those were acquitted by military judges.

Peers, the man who exposed the cover-up, was rewarded by having his career, which had been on the full-general track, completely derailed. He retired, still a lieutenant general, in 1972.

The tendency of bureaucrats to take a dim view of whistle-blowers such as Peers is particularly marked among those in the military. When Ernest Fitzgerald, a civilian employee of the air force, told Congress the truth about the C-5A in 1968, the air force fired him; when he won his job back through a court fight, the generals made sure he was given nothing to do.

The moral rot from Vietnam spread to the army as a whole. Even West Point, with its hallowed traditions of honor, did not escape. The early 1970s witnessed the worst cheating scandal in the history of the cadet corps, implicating hundreds instead of the handful of cadets who had been involved in such episodes in the past.

The army says it is concerned. Lieutenant General Andrew J. Goodpaster, superintendent at West Point, included morals and ethics in the academy's curriculum. "Under the new program," reported Drew Middleton in the August 6, 1978 issue of the *New York Times*, "plebes or first-year students must take two ethics courses. One course will be required in each subsequent year."

Goodpaster explained to Middleton that the new courses would teach the young officer how to act when his superior presses him for such things as fake body counts.

"All he has to do," Goodpaster said, "is to ask his senior officer, 'Sir, are you asking me to send in a fake report?' That will do it."

If you find that dialogue a little hard to believe, you may be fortified in your skepticism by the revelation in December 1978 by Herman Smith, former coach of Army's football team, that Goodpaster himself had participated in the cover-up of scandals about the recruiting of athletes for academy teams.

The aspect of moral blindness in the military that probably costs the taxpayers the most is the growing retirement custom under which thousands of officers leave the service each year to go to work for private companies that have contracts with the armed forces. Not only do we pay them two salaries—their pension and their salary from the contracts, which are usually financed by the U.S. Treasury—but they are tempted during their service careers to soft-pedal any criticism they might have of those military contractors, who are, after all, their potential retirement employers and who are therefore important parts of their survival networks. Consequently, the contractors are more likely to get away with building defective planes or tanks or ships, or, what is more common, with charging too much for them.

Another aspect of the retirement system that is unduly hard on the taxpayers: It permits retirement after only twenty years' service. The original reason for early retirement was that most military jobs in days gone by were physically arduous. If you had to spend twenty years charging over hills on foot—or even on horseback—you needed the promise of early retirement to help you endure. But now things have changed. Most military jobs are essentially deskbound and technical, clerical, or managerial in nature. In fact, in 1975 the General Accounting Office found that 93 percent of the enlisted personnel who were taking pensions after an average of twenty-one years of service had been working in support-type jobs that had none of the hardships of combat. So an early retirement designed to reward combat-weary troops is being used instead to benefit paperwork-weary bureaucrats.

The pension system is just part of a larger pattern of our fiscal indulgence of the military. For example, the taxpayers spend $350 million a year to subsidize commissaries where service families buy food and other items at a 25-percent discount—this while army officers receive up to $13,465 per year more in overall compensation and benefits than do federal civilian employees in comparable grades of the civil service.

All this overindulgence creates a psychology in which military personnel begin to think they have a right to take us to the cleaners. Stephen Lynton of the *Washington Post* recently took a look at the taxpaying habits of the members of the military living in the Washington, D.C., area, and found that 51.9 percent did not file local tax returns in the District of Columbia—or anywhere else. They may live in Washington, but they raise the make-believe principle to new heights by claiming "residence" in one of the states that exempts military personnel from taxation. This means that the same taxpayers who pay their salaries have to pay their taxes as well.

If you can't make up your mind whether to laugh or cry, cry. Realize that those people could easily give us another Pearl Harbor. At the same time realize that most of them are not really the selfish buffoons they appear to be. Change the institutional imperatives that govern their lives and they will change.

There are five simple steps to take: (1) eliminate the practices that interfere with the development of unit pride; (2) provide that no one can become an officer without beginning at the bottom of the enlisted ranks and working his way up—earning promotions and assignments to places like the service academies and the war colleges—solely on the basis of demonstrated competence; (3) restore candor to the process of evaluating both officers and enlisted men; (4) require at least thirty years' service before retirement; (5) draft the rich. The last is the most essential step of all. How can the upper classes expect the soldier or sailor to be a dedicated, unselfish patriot when they refuse to share his risk? Why should he die for the people who look down on him, who secretly think he is a sucker?

FAMOUS CASES IN COURT REPORTING
WITNESS PROTECTION SYSTEMS

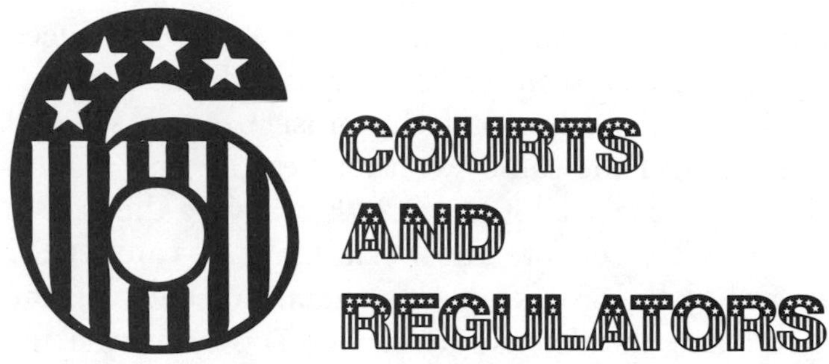

In 1971 a jury in New York City was split eleven to one in favor of a robbery conviction but was dismissed by the judge after only twelve hours of deliberation because he had another engagement. This enabled the robber to escape punishment because a second trial would have violated his constitutional rights. What was the other engagement? A trip to Europe.

The first thing to understand about our system of justice is that it is designed to serve the convenience of the people who staff it. If you have ever been a juror or a witness, you have probably wasted days, even weeks, waiting to be called while the judge, lawyers, and clerks confer on matters that may range from solemn to obscene. Whatever the purpose of these discussions, it is seldom to make the system suit the needs of the witness or the juror or, for that matter, of the plaintiff and the defendant—although on occasion, as in the case of the defendant in New York, one of the parties might be an unintentional beneficiary.

The cruelest result of this system can be the plea bargain, especially when a penniless criminal defendant is assigned a court-appointed lawyer who receives a fixed fee or salary regardless of how much time he spends on the case. His self-interest is served by minimizing the time and effort he devotes to his client's cause, and he is therefore tempted to persuade his client to plead guilty even though the client may have a good defense. And the lawyer has a good chance of getting the prosecution to accept a reduced charge, even if the prosecution is certain of the defendant's guilt. Again the reason is that the prosecutor—a salaried employee—is paid the same whether he tries the case or not. Trying the case is a lot of work. Not trying the case is *not* a lot of work. Therefore, most criminal cases are settled without trial by the process of plea bargaining between the prosecution and the defense attorney. Too often the innocent are convicted, and the guilty go free—all because lawyers don't want to work. Judges almost always approve plea bargains because they, too, are paid their salaries regardless of whether cases are tried, and they, too, want to get an early start on that fishing trip. The fishing trips, the ability to set their own schedule, may have something to do with why judgeships—particularly federal judgeships, which are well-paid, lifetime jobs—are looked upon as a reward. And when people look upon jobs as rewards rather than opportunities to work, there is a problem.

Most federal judges—at least most of those at the federal district court level—are appointed as a result of political activity on behalf of a U. S. senator. Because lawyers find it easier than most people to arrange their work schedules, they are able to do much more campaigning than the average person. Indeed, several federal judges once served as campaign managers for senators. As a general rule, presidents appoint federal district judges only upon nomination (or at least endorsement) by the senator from the district concerned. Jimmy Carter made a brief effort to change this practice when he became president, but by and large he was unsuccessful.

Donald Dale Jackson, in his book, *Judges*, quotes J. Sam Perry of Illinois on how he got his federal judgeship:

"I gambled," he said. "I saw a man—Paul Douglas—who looked as though he might be elected to the Senate. I backed him, and as a result I had his support. . . . I tried to obtain the appointment once before and learned that it requires not one but two senators. . . . I was out of politics and they did not need me. I decided I had better get back into politics, which I did. I learned that everyone shoots at the number-one choice, so I told each of the senators not to make me first. . . . That proved to be pretty good strategy—everybody else was shot off, and, no use lying about it, I helped to shoot them off. The result was I landed on top."

Generous campaign contributions also help, and even children can contribute, as in the case of Gilbert Merritt, a nominee for a judgeship on the Circuit Court of Appeals sponsored by Senator Jim Sasser of Tennessee. In response to a question by the Senate Judiciary Committee about the $1,000 contributions made by each of his three children, aged 10, 11, and 12, to Senator Sasser's campaign, Merritt said the contributions were the children's own idea: "The children were well aware of the need for campaign contributions, and were very well aware at that time of the need for such contributions by Senator Sasser."

This is not to say that these political appointees are without merit. On the contrary, most of them are quite able. (Their problem, remember, is the "reward" psychology that goes with the job.) The person who manages a senator's campaign is often a successful trial attorney with experience that is highly relevant to a district judgeship. In fact, for those judgeships the political appointee may be more qualified than the "merit" nominees of bar association panels. The latter are usually from the big law firms, where the training is less in courtroom skills than in the brief- and memo-writing skills that are more suitable for the higher federal courts—the Circuit Court of Appeals and the Supreme Court.

The men and women who become judges on these appellate courts are usually chosen because they earn credit for the administration that appoints them. This may be because they are supported by a powerful group, as Thurgood Marshall was backed by the NAACP, or because they enjoy wide respect within their profession, as did Lewis Powell, a former president of the American Bar Association.

Over the years it is the Powells who have dominated the appellate bench. Sometimes they come from the faculties of the great law schools (Felix Frankfurter), but more often they have spent most of their lives working for a major law firm in one of our larger cities. Most of them, whether they have come from faculty or firm, were "on the law review"—that is, they were staff members on their law schools' publications that analyzed recent decisions and trends in the law, jobs usually reserved for the best of the second- and third-year students.

These top students often go on to become law clerks for appellate judges. A great many of these judges were once clerks themselves, which may help explain why they have great respect for, and great dependence on, their clerks. Whatever the reason, these law clerks (distinguish them from the administrative clerks who serve as traffic managers and recorders of the court's business) have extraordinary power in the nation's legal system. In fact, they are the best examples I know of something I call "subordinate power," which is a powerful force in the life of most organizations. The person in charge wants to be liked by the people who work for him. And he wants them—at least those who are good—to continue working for him. Both of these factors mean that, just as the employee tries to please the boss, the boss will try to please the employee.

I can think of many times I have been saved from folly by employees who talked me into a more sensible course. I can also think of times when I've made mistakes because of a desire to please the people who worked with me. When I was working in the government, the director of my agency doubled my modest bureaucratic empire by adding a new division. I was eager to win the respect and affection of its employees, so I immediately endorsed their budget request for $1 million and worked hard to get it approved. Congress gave us only $500,000, but it turned out that we were able to spend only $385,000, even after using every end-of-the-fiscal-year spending technique known to the bureaucratic world.

So subordinate power can be harmful. You see its worst effects in the new cabinet secretary who brings to the job a determination to reform the department but quickly becomes the prisoner of his minions. The most powerful subordinates in America may be the thirty-two clerks who serve the justices of the Supreme Court. They advise and draft opinions for the justices. Sometimes they even write the opinion, as this anecdote about Thurgood Marshall, from Bob Woodward's and Scott Armstrong's *The Brethren,* reveals:

". . . a clerk once pointed out, 'You said that the right to privacy must go further than the home.'

'No,' Marshall retorted. He had never said that.

'Yes,' the clerk insisted.

'No, never,' Marshall was sure. 'Show me.'

The clerk brought the bound opinions.

Marshall read the relevant section.

'That's not my opinion. That's the opinion of a clerk from the prior term,' he declared."

The clerks also screen appeals. Consider the importance of just that function. In the 1930s, the Supreme Court received about 1,000 cases per year, and gave hearings and wrote opinions on about 150 of them. In 1978, the number of opinions was about the same (161) but the number of appeals had risen to 4,704, and the role of the clerks in choosing which could be heard had grown proportionately.

The increased number of cases coming before the Supreme Court is but one manifestation of explosive growth throughout our legal system. In one three-year period, 1969 to 1972, the case load for all the federal courts combined grew by 50 percent, and four times as many suits were filed in the state courts of California alone.

The law has a unique potential for growth. Every time a

lawyer files a suit he creates business not only for himself but for the lawyer or lawyers who will have to represent the other side. Americans have not been blind to this kind of opportunity: Between 1963 and 1982, the number of law students doubled to a total of 127,530. Motivation in the law, as in all human endeavor, is a murky matter, but three motives stand out: making money, making a name, and making law.

"Legal fees," wrote Professor Lawrence Tribe of Harvard Law School in 1979, "have soared to $25 billion a year." Money *is* important to lawyers. If you think "Deep Pocket" refers to some Watergate scandal, you're wrong. It's lawyer language to describe a client, usually a corporation, with an apparently limitless capacity to pay fees. When a lawyer has hooked one of these dream clients he can, as another elegant lawyerly phrase puts it, "keep the meter running."

The kind of case these lawyers love is exemplified by the one that arose out of the 1974 collapse of the Franklin National Bank. The lawyers for Michele Sindona, an Italian financier whose role in the bank's failure was such that he has strenuously resisted extradition, managed to con the presiding judge into disqualifying himself for having made a harmless joke about Sindona. The result was that the other clients in the case, who had already spent $1 million in legal fees, had to start all over again, while Mr. Sindona and all the lawyers laughed and laughed.

Perhaps a 1979 antitrust case came closest to revealing the relative rank of money and justice in the eyes of the legal profession. The plaintiffs' lawyers gave the court a 21-page memorandum on behalf of their clients' case, but to support their petition for $1.1 million in fees, they produced 200 neatly bound and carefully indexed pages.

Some lawyers want more than money. Take the late Abe Fortas, who made hundreds of thousands of dollars a year as a partner in one of Washington's leading firms, Arnold, Fortas, and Porter. But before he and his partners formed their firm, they had "made law" in the 1930s. Working for the very modest federal salaries of the time, they actually wrote much of the legislation of the New Deal. And even after he went into private practice, Fortas occasionally took cases that he thought were important even though his compensation for them was slight or nonexistent. In one of them, the Gideon case, he again made law by persuading the Supreme Court to rule that the Constitution requires defendants accused of felonies to be represented by a lawyer. However, it was the money motive that proved Fortas's undoing. Soon after he was nominated by Lyndon Johnson to be Chief Justice, it was revealed that he had continued to be a paid legal advisor to Louis Wolfson, who was a defendant in cases that could have gone to the Supreme Court, where Fortas could have ruled on them.

If Fortas's desire to make money got him into trouble, it was his desire to make law that got our legal system into trouble, both in terms of the regulatory law he and his partners pioneered in the 1930s and of the criminal law that was the outgrowth of his victory in the Gideon case. Let's look at the latter first.

Immortalized by Anthony Lewis of the *New York Times* in a book called *Gideon's Trumpet,* the case had every young lawyer who was eager to make a name filing writ after writ to free criminal defendants by making new law. Some of the new law represented clearly desirable humane reform. But the overall effect was to help the guilty escape punishment.

According to Ernst Van Der Haag, writing in *Policy Review,* only 20 percent of the 100,000 people arrested on felony charges in New York in 1976 were indicted by a grand jury. Far from all were actually convicted, and even fewer actually served time. Less than 2 percent of all reported felonies lead to imprisonment of the accused.

What in human terms do these statistics mean? Here is the story of the shooting of a girl named Sally Ann Morris as described by the *Washington Post:*

"She and her boyfriend, Henry Miller, were walking down 33rd Street, heading for an M Street restaurant . . . when two men approached. As they passed the couple, one of the men pulled out a gun, cocked it and stuck it in Sally Morris' back.

"Instinctively, Miller grabbed her and they started to run. . . . She heard gunfire and felt a slap at her back. . . . The bullet ripped through her intestinal tract and lodged in her lower abdomen. . . . Doctors had to perform a colostomy, rerouting the undamaged intestinal tract to a substitute opening in her lower abdomen. This type of operation allows body waste to be passed into a disposable plastic bag attached to the new opening.

"Police said the two men had been committing armed robberies in Georgetown for several weeks before the Morris shooting, escaping by hiding in the back seat of a getaway car driven by women.

"Compounding all this is the fear that the ordeal is not yet over and that her assailants may return to kill her. Four suspects arrested in the case, who were released on personal recognizance, pending trial, promptly disappeared and are at large today."

The Bail Reform Act of 1965, passed in the first flush of Gideon, applies to federal jurisdictions, including the District of Columbia. It provides, among other things, that "no financial condition shall be imposed to assure the safety of any other person or the country." In other words, bail could not be required to protect Sally Ann or the people of Washington from the man who had shot her.

What can we do, then, for the Sally Anns? The court's answer is "witness protection": She can be locked up for her own protection while the criminals who shot her go free until the trial. That really makes sense, doesn't it? Sally Ann in jail, the criminals on the street?

The law review writers, who want to make their names by making law, have over the years propagandized the federal courts into accepting insanity defenses that have freed one dangerous criminal after another. For example: Edward Carter, Jr., of Washington was charged with raping a 13-year-old girl but found not guilty by reason of insanity. Later, however, the doctors concluded that he had been lying to them when he convinced them he was insane. So the courts had to release him. He hadn't been convicted of the crime so he couldn't be imprisoned, and he wasn't insane so he couldn't be kept in a mental hospital.

If our criminal law does not protect society, neither does the other half of the Arnold, Fortas, and Porter legacy: regulatory law. A recent report on federal regulation made by a congressional oversight subcommittee concluded that the central characteristic of the nine regulatory agencies studied was their devotion "to the special interests of regulated industry and lack of sufficient concern for underrepresented interest," meaning the public. The report found that the Federal Power Commission and the Interstate Commerce Commission were the least effective of the nine agencies. The FPC, it noted, has "displayed a conscious indifference to the public beyond comparison with any other regulatory agency," the report said. As for the ICC, "Of all the agencies of government in Washington, there is probably no worse example of federal regulation than the Interstate Commerce Commission," writes Washington reporter Stephen Chapman. The ICC was created in 1887 to regulate the nation's rail traffic and was later given authority over waterways, pipelines, and highways as well. Chapman credits the commission with an almost spotless record of guarding and promoting the interests of the dominant companies and unions under its jurisdiction. It has pushed rates above their normal levels, inflated the costs of doing business, contrived to shut out newcomers, encouraged inefficiency, and made a thorough mess of surface transportation in the United States. No agency ever cried out louder for abolishment than the ICC.

"Perhaps the ICC's most important power today is its absolute control over entry into the trucking business,"

Chapman continues. In order to be admitted as a common carrier—essentially a commercial trucker—an applicant must obtain a certificate of public convenience and necessity, usually granted only when the applicant demonstrates that existing firms cannot provide the service specified. If established carriers show that they *can* handle the traffic, the application is normally turned down, even if the applicant could provide cheaper or more efficient service.

The difficulty of obtaining new certificates is suggested by their value on the market (the certificates, once obtained, can be sold to other firms without further ICC approval). Chapman tells of one company, Associated Transport Inc., that went bankrupt but was able to sell its operating rights for more than $20 million. The American Trucking Association says the price of operating rights quadrupled in a recent ten-year period and admits that they are a trucker's "single most important asset."

In fact, they may be more important than trucks. Many meat-hauling trucklines have dispensed with their employees, their tractors, and their trailers; now they simply rent their operating rights to independent truckers. The independent trucker does the work, but the firm with the operating rights gets a sizable cut of the take—generally 25 percent, but sometimes as much as 50 percent.

Trucking licenses are much like those the Federal Communications Commission issues to radio and television stations—only the FCC's may be even more valuable. Indeed, it is often said that they are licenses to print money. In the mid-1970s, the *Washington Post* sold its radio—not television, just radio—station in Washington for the staggering sum of $6 million. FCC licenses don't get much attention in the press, perhaps because so many of them are owned by large publishing operations.

American banks are regulated by three different agencies: 4,748 national chartered banks are regulated by the Comptroller of the Currency; 1,030 state-chartered banks that are members of the Federal Reserve system are regulated by the Federal Reserve Board; and the remaining 8,590 banks are regulated by the Federal Deposit Insurance Corporation. "Nowhere else," points out Senator Abraham Ribicoff, chairman of the Senate Governmental Affairs Committee, "does a regulated industry have an opportunity to select its regulatory agency."

A recent study by the committee cites the case of Bankers Trust Company of Columbia, South Carolina. It shifted from being FDIC-regulated to national status in 1973, acquired several smaller banks, and then shifted back to FDIC-regulated status fourteen months later. The FDIC had told the bank it would not approve the mergers, so, according to the study, the bank "switched charters, secured the Comptroller's approval of the mergers, and then switched back to the FDIC's jurisdiction."

To understand the regulatory mess, nothing is more important than remembering our principle that the system works less to serve the public's interest than the lawyers'. Take "the revolving door": A young lawyer just out of law school is hired by the Federal Trade Commission, where he gains expertise at the lower and middle levels. This makes him attractive to law firms that have clients with present or potential problems at the FTC. Once hired by the firm, he will usually stay in private practice, but he might return to the FTC for a few years at a high level, say as a director of the Bureau of Competition or as a commissioner. With that kind of experience, he can start his own law firm or name his own price with any of the major partnerships.

John Jenkins described a classic case in *The Washington Monthly*: "Federal Trade Commissioner Stephen A. Nye was in the job market. A San Francisco antitrust lawyer, Nye was cleaning out his desk at the FTC without any assurance, he told me, that a law firm would take him on. Had there been overtures from other firms? Not yet. It would be wrong to discuss employment with firms that had cases before the Commission. Nye would take a West Coast vacation, then start looking around.

"Shortly before Nye left the FTC last year, I examined his phone and appointment logs, as well as those of his fellow commissioners. Of the outsiders who were listed as having visited Nye to discuss Commission matters (a perfectly legal practice), the name of Wallace Adair stood out. Adair, an attorney with the Washington law firm of Howrey, Simon, Baker & Murchison, represented Kennecott Copper Corporation, which was under an FTC order to sell off its $1.2-billion Peabody Coal subsidiary. Kennecott had been dragging its feet. It wanted the FTC to reopen the case and amend its divestiture order, and had retained Adair in an effort to achieve that result.

"Weeks later, after Nye left the FTC, I phoned him at home. 'He's not here,' said the voice at the other end of the line. 'You can reach him at Howrey & Simon. He started there yesterday.'

"'He's probably in Mr. Adair's office,' the law firm's receptionist added cheerfully, when I called."

One consequence of the revolving door is that the large private law firm develops an expertise in manipulating the regulatory agencies, with the result that it usually succeeds in either defeating regulatory action against its clients or, at the least, in mitigating any unhappy consequences of those actions. Often delay will serve the client just as well as outright victory. And when the potential for delay in the courts is joined with that in the regulatory agencies, the possibilities are practically limitless.

There is, for example, the case of the El Paso Natural Gas Company. In 1959, after lengthy proceedings at the commission level, the Federal Power Commission approved the merger of El Paso with the Pacific Northwest Pipeline Company. The Justice Department appealed, however, and the case eventually made its way to the Supreme Court, which, in 1962, ruled that the FPC should not have approved the merger. El Paso's lawyers were masters at legal foot dragging, so it was not surprising to see in 1964 that the Supreme Court was once again ordering El Paso to divest Pacific "without delay."

Then El Paso's lawyers persuaded the federal judge who was overseeing the divestiture to adopt a plan so favorable to El Paso that outraged consumers brought the case to the Supreme Court once again, in 1967. The case returned to the Supreme Court two more times before finally being disposed of—and then only after the late Tommy Corcoran, the famous old New Dealer-turned-lobbyist, had made personal approaches to two justices, Black and Brennan, in an effort to get a rehearing for El Paso.

Why would a lawyer of Corcoran's standing engage in such scandalous attempts to lobby Supreme Court justices? The reason may well have been the large amount of money involved. Certainly money was the motive behind one major Washington firm's advice to a potential client with an antitrust problem. The story, as told to me by a lawyer who has worked for the firm, was that the client was told his legal situation was hopeless—he was doomed to lose the case—but the firm could stretch matters out for ten years or so, meaning the client could go on making money doing whatever he was doing wrong for those ten years. The client was asked whether he would be willing to pay $500,000 to $1 million a year for the legal fees that delay would require. His answer: "Of course."

Here was a client who knew he was wrong and a law firm that knew it was wrong. Yet they were both willing to delay the triumph of the public interest for ten years. That may have been bad for the public, but what about the $5 million to $10 million it would have cost the client? Wouldn't that hurt? Hardly—legal fees are tax deductible.

Robert M. Kaus of *The Washington Monthly* once asked an attorney for the Environmental Protection Agency for an example of an average delay. He was told about a case in which the EPA spent two years haggling with the state of Idaho over the amount of sulfur dioxide a steel mill was releasing into the atmosphere. After taking testimony from every scientist and engineer in sight, the EPA ruled that 82 percent of the sulfur dioxide had to go. The state thereupon took the case to federal court, which after another two years' wait sent it back to the EPA so it could start all over again. The EPA lawyer said, "I just think that's life. I don't think it's something that has to be reformed."

Because lawyers make money out of the system as it is, even those who advocate reform usually want to add legal procedures. The most ironic reform idea of all is Ralph Nader's proposal to establish a Consumer Protection Agency, whose lawyers would intervene on behalf of the public in proceedings of the existing regulatory agencies—the same agencies that were originally established to protect the public.

In a 1974 book called *The Genteel Populists,* Simon Lazarus, a former public interest lawyer who went on to join Arnold and Porter and then the White House staff, concludes that the real hope of regulatory reform lies in making the regulators' decisions more easily appealable to the courts. You have to be a true believer in the adversary system to think more of it is the answer. In fact, the adversary system is itself central to the problem. What it does is pit two sides against one another, with self-interest motivating the lawyers less toward the pursuit of truth and justice than toward the pursuit of victory.

The way lawyers view truth and justice is suggested by this advice from the book *How to Cross-Examine Witnesses Successfully*: "No matter how clear, how logical, how concise, or how honest a witness may be or make his testimony appear, there is always some way, if you are ingenious enough, to cast suspicion on it, to weaken its effect."

Ann Strick has described the American system of justice as a modern version of the medieval trial by battle, in which the strongest won. The richest win now because they can hire the brightest lawyers. Between the lawyers stands a judge, who in the main tradition of American law is not a seeker after truth and justice but a referee who seeks only to ensure that the combatants obey the rule of fair play. This means, among other things, that he does not intervene to protect the poor man with a bad lawyer from the rich man with a good lawyer.

It is possible for disputes to be handled otherwise—by judges who seek justice and truth, as do many English jurists and as did Judge Sirica in the Watergate case; who keep lawyers out of the courtroom and take responsibility for protecting each party's rights, as do some small claims courts. It is also possible for disputes to be settled by mediators who seek solutions that are fair to both sides and that will enable the parties to leave the dispute as friends rather than enemies.

The adversary system is cruelest in divorce cases. Parents who, for the sake of their children, need to go forward in life with mutual respect and affection, are encouraged to blame one another for what went wrong in their marriage and often leave the courtroom with hatred in their hearts.

The auto accident is the most absurd of all adversary proceedings. In most cases, what happened was an *accident.* If anyone did it deliberately or recklessly, he could be prosecuted as a criminal. But, with no one really to blame, or with both parties sharing responsibility, the adversary system says you have to prove that it was all the other guy's fault.

These auto accident cases inflate insurance premiums for everyone and cause endless delays in payment to the injured. For this reason sixteen states have adopted no-fault systems for settling personal injury claims arising from auto accidents. But an attempt to get such a system enacted nationally was defeated in the Senate in 1976: Intensive lobbying by the American Trial Lawyers Association (the fellows the senators need at campaign time) brought it down.

In China, disputes of this type (there they usually involve bicycles) are settled by mediation. Our own government used to have experimental Neighborhood Justice Centers in three cities, where mediators handled consumer, landlord, employer, and family problems. Bearing in mind the time the regular legal system takes to handle disputes, it is interesting to note that hearings at the Kansas City center usually took just two hours. And it took only thirteen days (on the average) from the time the case was filed until it was heard. In their inimitable way, the Reagan budget cutters did away with the program.

If the government really wanted to save money and unclog the courts, it would expand rather than shut down such experiments and do two other things. First, as Jonathan Alter has suggested, litigants should be charged a "user's fee" in civil cases (such as contract disputes) where wealthy private parties are arguing over exclusively private matters. Since 90 percent of us never use the courts in our entire lives, why are we paying the full cost of putting on trials for wealthy adversaries?

The second reform is to adopt a "loser pays" provision. In California, the legislature now provides for mandatory arbitration of all cases involving amounts less than $15,000. If either party is unhappy with the arbitrator's decision, the case can go to trial, but the loser can be ordered to pay the other party's costs for both the trial and the arbitration.

In England, the courts can compel the losing party to pay the winner's legal costs. Probably no single reform would do more to reduce the amount of litigation than the adoption of this English rule for all American courts. Then only those who were confident of the rightness of their cause would go to court.

Why do most American lawyers oppose adopting the English rule? It would mean less business, which brings us back to where we started: The American legal system serves the lawyers, not the people.

IN·GOD·WE·TRUST
BALL BEARINGS

CONGRESS

The most striking feature of a congressman's life is its hectic jumble of votes, meetings, appointments, and visits from folks from back home who just drop by. From an 8 A.M. breakfast conference with a group of union leaders, a typical morning will take him to his office around 9, where the waiting room is filled with people who want to see him. (If the congressman is unlucky enough to be new and assigned to the overcrowded Longworth Building, the waiting room will also function as an office for as many as ten members of his staff. The overall effect can only be compared to the Marx Brothers' cabin in *A Night at the Opera.*)

From 9 until 10:30 or so, he will try to give the impression that he is devoting his entire attention to a businessman from his state with a tax problem; to a delegation protesting their town's loss of air or rail service; to a constituent and his three children, who are in town for the day and want to say hello; and to a couple of staff members whose morale will collapse if they don't have five minutes alone to go over essential business with him. As he strives to project one-on-one sincerity to all these people, he is fielding phone calls at the rate of one every five minutes and checking a press release that has to get out in time to make the afternoon papers in his district.

He leaves this madhouse to go to a committee meeting, accompanied by his legislative aide, who tries to brief him on the business before the committee along the way. The meeting started at 10, so he struggles to catch the thread of questioning, while a committee staff member whispers in his ear. And so the day continues.

Ironically, despite all this activity, most members of Congress feel a sense of isolation most of the time. The people they see—constituents, lobbyists, reporters—usually want something, which means the congressman must be guarded in his relationship with them. And there is less relaxed comradeship with fellow members than there used to be. In the days of short congressional sessions, members left their families back home and spent their evenings together. The House Ways and Means committee used to have dinner one night a week at a restaurant in downtown Washington. These were also the days of unrecorded legislative action, when much of the voting on the floor was done in ways that protected the public's right not to know where its representatives stood. Back then many a congressional friendship was cemented by an exchange of votes. Such mutual favors are possible today only when the recorded vote is certain not to damage either congressman's standing with his constituents.

A member of Congress is probably closest to his staff. They truly care about him, at least in the sense that their jobs and futures depend on his performance and eventual re-election. But this identity of interest does not relieve him of anxiety about what they, should they someday choose to leave his employ, might say about what he has said or done in an unguarded moment.

This is the main reason why politicians have few close relationships. As Ross Baker of Rutgers has pointed out, "The establishment of a personal friendship involves a degree of risk, for . . . there is the likelihood of self revelation, the exposure of innermost thoughts, the exchange of confidences, the laying bare of personal problems."

The arrangement of offices on Capitol Hill tends to reinforce a member's sense of isolation. Each senator or congressman, with his staff, occupies a suite of offices set off from his colleagues. (Although his case may have been extreme, it is said that the late John McClellan, a senator from Arkansas for over thirty years, never set foot in the office of another senator.) To be sure, other members are encountered in committee meetings and on the floor, but in each place the focus is necessarily on the business at hand. They can meet in the Senate or House dining rooms, but there one has the feeling that everyone is "on" in the theatrical sense of the word: performing and posturing; wondering who that familiar and possibly famous person is with Senator Y; asking himself whether he should rise to greet the National Committeeman who is passing by; and often, even after he has been a member for years, being almost as awed by his surroundings as the average visitor.

More often than not, when first elected, the member of the House was not a prominent citizen. Often, he had reached his mid-thirties without having achieved conspicuous success and was ready to gamble to make it, to pursue the long shot of defeating an incumbent congressman. He did not run because he had a program. He ran because he wanted a good job. The greatest indicator of how unexciting most congressmen's lives were back home—of how truly uninvolved they had been in their own communities—is that very few return home when the day of their own defeat comes. But if he ran a particularly effective campaign, or ran against a particularly ineffective opponent, or happened to run in a "tidal wave" year (1934, 1946, 1958, and 1964, for instance), he won the gamble. Having kept his spirits up during the campaign with thoughts like, "Well, even if I lose, the publicity will help my law practice," he now finds himself a congressman.

Usually a newly elected congressman is so impressed by his opponent's misfortune that his first resolution is to make sure a similar tragedy does not befall him. And this helps explain the fundamental first rule of life on the Hill: Each member is concerned above all else with getting re-elected. Nothing will help you understand Congress better than the re-election imperative. Do you wonder why the average staff member spends four-fifths of his time on constituent service and one-fifth on legislation? Are you puzzled by the fact that Congress does little to remedy those defects in the bureaucracy against which it constantly rails? The reason is that the Congress itself has a stake in bureaucratic ineptitude.

As Morris Fiorina of the University of California has pointed out, the more bureaucrats do wrong to the public, the more favors congressmen can do for their constituents as they right the wrongs—or as they appear to try to right them. *Appear to* is the key to understanding the congressional world of make believe.

Suppose that your local postmaster spends his time reading his mail rather than seeing that yours is delivered, and the poor mail service is hurting your business. You're mad as hell, and you write your congressman to say you're not going to take it anymore. He promptly and efficiently forwards your complaint to the Postal Service, which responds that nothing can be done because your postmaster is protected by civil service laws. Your congressman then promptly forwards you a copy of the Postal Service's letter, and puts you on his mailing list—after which you receive (allowing ten days for delivery) regular copies of the Congressional Record, containing the Congressman's denunciations of postal inefficiency. "Well, I guess nothing can be done," you sigh, "but my congressman certainly seems to have done his best," and you then resolve to vote for him next time around.

Getting re-elected, in addition to taking care of—or appearing to take care of—specific constituents, also involves avoiding any action that might antagonize a sizable interest group. Since most members have nightmares about losing by ten votes, a group with eleven members can have considerable influence. This is the basic reason why lobbies run the American government today.

The growing power of the lobbies is best illustrated by

the growing tendency of Congress to organize itself into interest-group caucuses such as the well-known Black Caucus. But have you heard of the Suburban Caucus? There is one. There is also a Steel Caucus, and a Ball Bearing Caucus. My favorite is the Mushroom Caucus.

Interest groups have always been with us, but they used to operate within coalitions called the Democratic and Republican Parties. As long as these groups functioned inside a party, they were compelled to compromise with the other groups under its umbrella. But, as first federal and then state and local governments embraced civil service and deprived the parties of the patronage that had glued them together, the parties declined and the power and the assertiveness of the lobbies grew—along with the ability to enforce their demands through campaign contributions.

It's easier for a member than a challenger to get contributions. The lobbyist knows that the odds favor incumbents. Congress has helped make the odds by enacting measure after measure designed to keep the incumbent in office. For instance, incumbents have access to television facilities in the capital where they can tape programs for stations in their district. They also have free mailing privileges. Each time a member sends a newsletter to his constituents, he not only has the staff's work in preparing it paid by the taxpayer, but, assuming a typical mailing of 200,000, gets a postal subsidy of from $16,800 to $40,000, depending on whether you think first or third class rates should apply.

Congress takes great pains to avoid reminding the voters that they are financing such mailings. The postage-free (franked) envelopes used for mailings usually bear only the member's signature. Recently, however, a staff error resulted in the appearance of the phrase "Postage Paid by Congress" under the facsimile signature. But Representative Morris Udall reassured his colleagues:

"I'm happy to report that the House committee on the

Post Office and Civil Service today added a repealer of this unfortunate goof to a Senate-passed bill, S. 2315, and that the bill is expected to be on the next Consent Calendar.

"Members who are concerned about using envelopes with the new language might consider withholding orders for new supplies of franked envelopes for a few days until the repealer can be signed into law.

"I personally regret this error and any difficulty it might have caused."

S. 2315, as amended, passed the House by unanimous consent, and the Senate passed it two weeks later. It is reassuring to know that our Congress is capable of rapid action in genuine emergencies.

The ingenuity of congressmen in finding ways to tap the public till in the interest of their own re-election is perhaps best illustrated by this editorial from the *Washington Post* about former congressman Ted Risenhoover and Flag Day:

"The Risenhoover campaign committee has bought television time to air . . . a half-hour version of the official Air Force film of this year's Flag Day observance in the House. As edited by Rep. Risenhoover at his own expense (a modest $650, compared with the $5000-or-so the Air Force spent to produce the film), this epic documentary includes an inspirational address by evangelist Oral Roberts, whose headquarters is in Tulsa. (Risenhoover is from Oklahoma.) But the dramatic climax is a reading of 'I Am an American' by the chairman of the House Flag Day Committee, backed up by the Air Force's Singing Sergeants and band. Do we have to tell you who is this year's chairman of the House Flag Day Committee?"

After wondering how it came to pass that the Air Force had filmed this minor event, the *Post* continued:

". . . it was not a one-time favor to a two-term congressman. Instead it has become routine over the years for the services to supply a band and a camera crew for the House's annual Flag Day show. The Air Force simply got the call this year.

"Does this mean there's an intense nationwide demand for—and use of—Flag Day movies as inspirational programs? . . . Well, not exactly. The film is quietly handed over to the Flag Day chairman, and his office distributes it. Mr. Risenhoover thinks highly of this service; he shared last year's movie—in which he was shown making a patriotic speech—with some 200 groups in his district. Apparently no one else used it."

Does the congressman or senator, you must be beginning to wonder, ever spend any time on legislation? Yes, he does, and we're coming to that. But first you should realize that his concern with legislation is often less with its substance than with its potential impact on his campaign for re-election. There is no better illustration of this than a resolution guided by Representative Romano L. Mazzoli through the 92nd Congress Democratic Club. It read:

"WHEREAS, in the 92nd Congress the Democratic Party enjoys a majority of the membership of the House of Representatives; and

"WHEREAS, it is in the best interest of the people of the United States that this majority be maintained or expanded in future elections; and

"WHEREAS, the development of a substantial legislative record is a major assistance in re-election campaigns;

"Be It Resolved By The 92nd Congress Democratic Club:

"That the Democratic Leadership of the House of Representatives of the 92nd Congress is requested and urged to develop a program whereby first-term Democratic Members of Congress receive advance notice of important legislation likely to receive Party support and thus, likely to become law, and that such interested Members be afforded an opportunity to join in the sponsorship of such legislation in order to compile a substantial legislative record on which to run for re-election."

Before we get to the members' work on legislation, we should note other main uses of their time. Part of the motivation in seeking office was the chance to lead the life they imagined was led by important men in Washington. This involves lunches in fancy restaurants with celebrities, weekends at sporting events and hunting lodges, dinners with the social and financial elite, and in general just being with people who give a warm sense of importance. And then there is the traveling—trips at the taxpayers' expense to such essential events as the Interparliamentary Union, which usually manages to hold its meetings in places like Geneva or Lisbon.

The late Marvella Bayh's autobiography makes it clear that she and her husband, Birch, the former senator from Indiana, enjoyed the pleasures of Interparliamentary travel three times in just one ten-month period. The grandest was a three-week "study trip" to seven Asian countries that took place in December 1969, but there was also an April 1970 jaunt to Monaco and an excursion to Paris in September of that year. Indeed, Mrs. Bayh recalls her first words on hearing from her father that Birch had been elected to the Senate: "Daddy, do you know what this means? It means some day I may go to Europe!"

The champion traveler of them all was William L. Scott, who served as senator from Virginia from 1972 to 1978. He managed in that time to get to Afghanistan, Australia, Colombia, Peru, Brazil, Argentina, Panama, the Philippines, Belgium, West Germany, Spain, Turkey, Romania, India, France, Greece, Italy, England, Japan, Hong Kong, Cambodia, Vietnam, Thailand, Taiwan, Indonesia, Saudi Arabia, Jordan, Syria, and Austria.

Congress also offers other forms of recreation for its members. The powerful have hideaway offices that are not by any means used exclusively for business. There are saunas, gymnasiums, and swimming pools and, if alcoholic refreshment is desired, free drinks are available in the offices of the Secretary of the Senate and certain other congressional officials. And, of course, there is free parking.

In Washington, parking costs the average person $90 per month. So, when Senator Gary Hart proposed in 1977 that congressmen and their staffs pay from $10 to $50 per month, depending on their incomes, it seemed the most modest of reforms. The congressmen would still be getting a bargain. The small charge, it was assumed, would come out of the substantial salary increases granted that year to congressmen and their employees. If anyone didn't want to pay he could use public transportation and save energy. All in all, an irresistible case.

When the vote came in the Senate, however, Hart lost (65 to 28). Note among the yeas (those voting *for* free parking) the number of conservatives, who say they oppose government handouts, and of environmentalists, who supposedly support energy conservation:

Yeas, 65

Anderson	Domenici	Javits	Sarbanes
Baker	Durkin	Laxalt	Sasser
Bartlett	Eagleton	Long	Schmitt
Bayh	Eastland	Lugar	Schweiker
Bentsen	Ford	Magnuson	Scott
Brooke	Garn	Matsunaga	Sparkman
Bumpers	Glenn	McClellan	Stennis
Burdick	Gravel	McIntyre	Stevens
Byrd, Robt. C.	Griffin	Melcher	Stone
Cannon	Hansen	Metcalf	Talmadge
Case	Hathaway	Moynihan	Tower
Chiles	Hayakawa	Muskie	Wallop
Church	Hollings	Nelson	Williams
Cranston	Huddleston	Pearson	Young
Curtis	Humphrey	Randolph	
DeConcini	Inouye	Ribicoff	
Dole	Jackson	Riegle	

Nays, 28

Abourezk	Clark	Heinz	Percy
Allen	Culver	Kennedy	Proxmire
Bellmon	Danforth	McGovern	Roth
Biden	Goldwater	Metzenbaum	Stevenson
Byrd, Harry F., Jr.	Hart	Nunn	Thurmond
Helms	Haskell	Packwood	Weicker.
Chafee	Hatfield	Pell	Zorinsky

But Congress does do important work. Enough of its members combine idealism and intelligence—there *were*, after all, 28 Nays on that parking vote—to make them and their generally younger and even more idealistic staffs the most impressive pool of talent in Washington. They take their responsibilities seriously. These duties include determining the need for legislation, writing the laws, and investigating how they are carried out.

Sometimes there is conflict between these investigaive and legislative functions. Congressman A, while diligently pursuing investigative leads, may find himself exposing defects in Congressman B's favorite program. This means B will be less likely to support bills sponsored by A.

Bills are introduced by members on their own initiative or at the suggestion of the executive branch or of a lobbyist for the private sector. Once introduced they are referred to committees, which is where most of the substantive work of Congress takes place. In 1979 the House had 152 committees and subcommittees—which means among other things that 152 of the 270 Democrats then in the House could be chairmen and enjoy the power and publicity attendant to that role. The largest committees were Banking, with 121 staff members; Interstate and Foreign Commerce, with 115; Education and Labor, with 111; International Relations, with 105; and Ways and Means, with 92. Except for Rules, Ways and Means, and Appropriations, which are exclusive assignments, a House member may serve on one major committee and on one other less important committee. In the Senate, where the committee load is heavier, a typical member could be on three committees and six subcommittees.

Until the trend was arrested in 1981, the size of committee staffs had been increasing dramatically in recent years. In the House, for example, there were 634 people on the committee staffs in 1967. By 1980 that figure had more than tripled. The main reason for this growth was that Congress no longer trusted the executive branch to provide it with accurate information. Vietnam and Watergate combined to make people on the Hill wary of what they were being told by the people downtown. This has meant hiring more congressional staff to check the facts alleged by the administration and to search for the unhappy news that the executive branch so often omits from its reports. In this work Congress is aided by its investigative arm, the General Accounting Office, an organization that has responded to the lessons of Vietnam and Watergate by developing an investigative potential considerably beyond the fiscal post-audit function to which it was once largely confined. Still, like its fellow investigators from the press and the congressional committees, the GAO tends to spend more time looking for fraud and illegality than in appraising program effectiveness—whether the programs Congress enacted actually work, and if not, why—the information Congress needs most. And GAO reports tend to be dull reading, which of course limits their audience and their effectiveness.

Perhaps the most useful congressional reform of recent years has been the establishment of the Congressional Budget Office and the House and Senate Budget Committees. Before these committees existed, Congress had no way of evaluating the budget priorities given by the executive branch. Furthermore, it had no effective way to discipline itself on expenditures. Now, by budget resolutions, it establishes targets in May and final ceilings in September. The Congressional Budget Office gives Congress a way of evaluating executive branch assertions that this weapons program will cost X dollars or that tax measure will produce Y dollars.

In another reform motivated by Vietnam, Congress passed the War Powers Act of 1973. This was the first time in history that it had defined and limited the president's power to make war.

These measures, however, do not begin to make up for the powers that Congress has lost in the last half century. Congress itself has been largely responsible for this decline, through its enthusiastic adoption of two "reforms"—

delegation of rulemaking power, and revenue sharing. Delegation became popular during the New Deal, when Congress got in the habit of setting up agencies to solve problems, while giving them only the vaguest of instructions. So the National Labor Relations Board was established to ban "unfair labor practices"; the Securities and Exchange Commission, to prohibit "manipulative or deceptive devices" in the sale of stocks and bonds. When the agencies wrote "rules" in an attempt to give substance to these platitudes, the rules had the force of law—except that they were laws Congress had never voted on. In effect, Congress repeatedly gave away a sizable chunk of its power to the unelected civil servants who staff the federal agencies.

Revenue sharing is a more recent innovation. As initiated by President Nixon, the idea was to transfer power away from the unaccountable federal agency officials to state and local officials who were presumably closer to the people. In fact, however, revenue sharing—and the spectacular growth of other federal subsidies to state and local governments—has given those governments the wherewithal to finance their own imitations of federal agencies and the federal civil service. So the effect has once again been to shift power away from a potentially accountable Congress to unaccountable bureaucrats—although this time the bureaucrats are in the state capitals and city halls rather than in Washington.

As a result, Congress today is largely a reactive body. The Founding Fathers thought that Congress would make policy and that the president would execute it. The actual process, however, is that a lobby or the administration proposes; Congress assents, amends, or rejects; and then the executive branch and the states "implement."

In many areas where congressional assent is necessary, it is almost automatic. The last time a cabinet nominee was turned down by the Senate was in 1959. As one staff member told Susanna McBee of the *Washington Post*: "They can be crummy, mediocre, not qualified, even in industry's pocket, and if they haven't done anything criminal, they're approved. You almost have to be found with one finger in the cookie jar to get rejected."

Even when Congress *wants* to carry out its duties, there are obstacles. One is the way it is organized. For example, responsibility for federal pension programs is divided among eleven committees of the House and ten of the Senate. Jurisdiction over energy legislation is even more fragmented.

Another obstacle to congressional effectiveness is the communication gap between the executive and legislative branches. The culture of the Hill is vastly different from the culture of bureaucracy. While both cultures place a high value on holding onto one's job—on surviving in Washington—the congressman who has to risk all on the throw-of-the-dice of elective politics does not easily empathize with the bureaucrat who seldom has to risk anything. And the bureaucrat, of course, regards the congressman as a "politician," which, in the lexicon of the civil service, is removed by only the most delicate shade of meaning from "crook."

The fact that the bureaucrat and the congressman work in different cultures becomes most evident when an executive branch witness is appearing before a congressional committee. Unless he has had some political experience—and usually this is true of only a few people at the top of an agency—he is almost certain to be defensive and motivated primarily by the desire to protect himself and his agency. He is not forthright. Indeed, when I worked in the executive branch I felt it was my duty to conceal from Congress any fact that might reflect adversely on my agency. The congressmen, on the other hand, were usually ill-prepared (remember what their typical day is like) and seldom asked me the right questions. When they did, it appeared to be accidental, and they failed to ask the right follow-up questions.

Congress seldom requires executive branch witnesses

to testify under oath. Ninety percent of such testimony is not under oath, and 90 percent of it is removed from precise truth by degrees varying from the subtlest unconscious nuance to the grossest and most deliberate distortion. The danger of all this is that, unless ways can be found to remind executive branch witnesses that they and the Congress work for the same United States and that it is not in the interest of the country for there to be a lack of candor between the branches of its government, it is possible that we will end up with another giant bureaucracy on the Hill to check on the one downtown.

We have already discussed the growth of committee staffs. The same thing has happened to the staffs of individual congressmen. In 1979, Courtney Pace retired after thirty-five years as a staff member for former Senator James Eastland. When Pace first joined there were but three other people on the staff and the total payroll for all was $11,000. When he retired, there were twenty-four people working in Eastland's office in Washington and eight or nine more in his home-state offices. The payroll had grown to $525,000. Members of the House have staffs almost as large. Each, regardless of seniority or committee assignment, is allotted eighteen employees.

Staff members have little interest in blowing the whistle on this particular situation. Instead, they are dedicated to convincing their boss that their work is essential. So, just as the caseworker—the staff member who handles constituent problems—doesn't want agency X reformed because its misdeeds supply half his work, the legislative assistant doesn't want the congressman to have other sources of information about legislation. Every staff member is tempted to be an Iago contributing to the member's isolation.

One of the great truths of Washington life little known by the folks back home is the power of congressional staff members. Norman Dicks, who was elected to the House in 1976 after spending eight years as an aide to Senator Warren

Magnuson, was recently quoted by Martin Tolchin of the *New York Times* as saying, "People asked me how I felt about being elected to Congress, and I told them I never thought I'd give up that much power voluntarily." When Laurence Woodworth, who had been staff director of the Joint Committee on Internal Revenue for fourteen years, died in 1977, Representative Al Ullman said, "In his quaint way he was as much an influence in shaping tax policy in this country as any committee chairman or treasury secretary or president in recent memory." Similarly, until he resigned in 1976, Harley Dirks of the House Appropriations Committee staff was known as the man to see on anything affecting labor, health, education, or welfare appropriations. And Richard Perle, when he was assistant to Senator Henry Jackson, was the most influential force on Capitol Hill on arms control and disarmament issues.

Congressional staffs are, like Supreme Court clerks, another great example of subordinate power. No one knows this better than the lobbyists, who in a recent survey rated the congressional staffs as their number one lobbying target (by contrast, the White House ranked sixth). The member himself is well aware of this power. He wants bright, aggressive people on his staff. This means that, in order to keep them satisfied, he will have to, at least occasionally, introduce the bills they want enacted and ask the questions they want raised. Senator John Culver explained to Elizabeth Drew:

". . . You get a bright staff person who works for months on something in the subcommittee that he's particularly interested in, and finally you don't want to disappoint him or her, and so you say, 'Go ahead,' only to regret it later because you find yourself involved in something that you don't have sufficient interest in, and spending your energy and political capital on frustrating and unsatisfying efforts."

The congressman can become too dependent on his staff. There was, for example, the time Senator James Sasser lost his place in a statement he was making to the Government Affairs Committee. His aide was, unfortunately, seated several feet away, so everyone could hear Sasser's agonized whisper, "What comes next?"

This dependence, if the staff work is bad, can leave the congressman out on a very long limb, as was illustrated in 1977 by Senator Charles Percy's questions about Bert Lance's use of his company plane. Percy, with a choice of over 100 trips by Lance to his vacation home at Sea Island, asked about one on February 5, to which Lance was able to reply, "That was to the American Banking Association convention that was held at Sea Island."

Then Percy questioned a Lance trip to Warm Springs, asking if the nearest airport to Warm Springs wasn't at Lynchburg, Virginia. Lance pointed out that to Georgians, Warm Springs means the one in Georgia, not the one in Virginia.

Congressional wives take delight in pointing out that kind of error to their husbands. Hostility bordering on open warfare is typical of spouse–staff relations. The wives who live through their husbands—they are less common in this era of liberated women, but they still exist—often secretly think of themselves as "the real senator." Since the husband's administrative assistant—or "AA" as the staff member is called on the Hill—usually thinks he is the real senator, conflict is practically inevitable. On the other hand, the person most victimized by the congressman's life is the wife who seeks an honest relationship of mutual support and respect.

The typical day that opened this chapter usually ends around 11:30 P.M., as the congressman leaves an embassy party, at which he has been hustling as if it were a key precinct on election eve. He is too tired to talk about any but the most trivial matters, too tired usually to do anything but fall into bed and go to sleep. This is why most congressional marital disputes—at least the really serious ones—are faced only on vacations. As Marvella Bayh wrote:

"All those bottled-up complaints most people deal with and work through on a day-to-day basis suddenly surface. For us, there had been no day-to-day basis in the six years we had been in Washington, except for vacations. Then the bottle came uncorked."

Except for such rare occasions, then, the congressman's isolation is complete. He doesn't even communicate with his wife. He shakes hands with hundreds of people every day, but he really talks to no one.

This lack of real relationships with other human beings is typical of the unreality of Capitol Hill. If the reporters covering the Congress are wrapped in the cocoon we described in Chapter 2, the congressmen themselves are even more insulated. The Capitol and its six satellite office buildings constitute a self-contained world connected by underground passageways that permit the congressmen, without having to encounter life outside, to enjoy the services of barbers, nurses, credit unions, travel bureaus, cafeterias, and restaurants. In 1979 one congressman managed to turn his office into an apartment and lived there around the clock.

It is easy for those enveloped in this cocoon to imagine that *it* is the real world—to think, for example, that they are affecting reality by enacting legislation when all they are really doing is passing a bill that may, depending on how it is carried out by the administration, have no effect at all. This is the ultimate make believe.

The make believe could be eliminated if Congress systematically followed its laws through the bureaucracy to see what finally happened. Did the law solve the problem it was designed to meet? Or should it be repealed or amended in light of experience? This kind of follow-up, ironically called "legislative oversight" in the jargon of Congress, is one of the most neglected of congressional duties. The reason for the neglect is simple: There are no votes in oversight. The people simply do not understand its importance.

Even when there is oversight, it is likely to be perfunctory. Often, this is because survival networks are involved, and the committee chairman does not want to be hard on an old friend. Even when survival networks are not a factor, key committee members may perceive an identity of interest with the executive department concerned. For example, the chairman whose district's prosperity depends on defense contracts is not likely to be a severe cross-examiner of witnesses from the Department of Defense. He knows his own survival at the polls would be endangered by excessive zeal as an overseer, zeal that might anger the DOD and cause cancellation of his district's contracts. In other words, the average congressman knows that effective oversight may lose votes. Since he also knows of no sign, not even the faintest indication, that the public understands the importance of skillful oversight and might reward him at the polls for it, the ultimate villain behind make believe on Capitol Hill is the ignorance of the people and the sloth and ineptitude of those who are charged with informing and educating them.

HOTTER
CHAUDE

THE PRESIDENCY

A president can also be insulated from reality. The White House, Camp David, and Air Force One are even more protected than Capitol Hill. This insulation can be compounded by a president's personal characteristics. Ronald Reagan came to the White House having spent the preceding 20 years of his life surrounded by rich Californians and having filtered out the harsh memories of his childhood and youth during the Depression so that the real world for him had become one part Rodeo Drive and one part Norman Rockwell.

Jimmy Carter celebrated his first Christmas in office by giving a reception for the members of the White House staff and their families. But he provided nothing to eat or drink—not even a potato chip or a Coke—and to top it off, neither he nor any other member of the first family bothered to show up. As this story suggests, Carter did not have a warm personal relationship with most of his staff. Only a handful saw him socially or, more importantly, on any business other than that which fit inside the stated duties of their positions on the organization charts. He talked to his speechwriters only about speeches, to his economics advisor only about economics, and even in those cases communication was largely in writing. His employees were not encouraged to make suggestions outside their nominal subject areas; he did not invite assistants to argue cases before him, hearing the strengths and weaknesses of each side's views. Instead, he spent long hours alone, reading memoranda, checking boxes to indicate what action should be taken. Within each department or agency, his realm of personal contact was usually limited to its head, and perhaps one or two others. He had not chosen the people just below the cabinet level—those who actually have their hands on the controls of much of the government—and maintained few contacts with them. This isolation meant that he heard very little dissent, that he seldom received the critical stimulation lively oral argument can provide, and that he had the personal loyalty of astonishingly few members of his administration.

What difference did this make to the country? For one thing, it meant bad decisions, because there was no one to point out to him the holes in what was proposed by the advisor with responsibility over whatever issue was at hand. The first example was the clearest—the energy program that James Schlesinger gave President Carter in April 1977. Reagan made the same mistake in April 1981 when he accepted proposals for social security reform that had not been adequately argued out by his staff. The result was a panic among the elderly that has almost certainly set back the cause of real social security reform for years.

After the energy disaster, Carter permitted all officials whose interests might be affected by a proposal to make written comments before it landed on his desk. But there was still little face-to-face argument in front of or with Carter. And there was still no involvement at all of the bright people on issues that might be completely outside their sphere of formal responsibility but about which they might have something helpful to say. The result was policies that were full of holes, internal contradictions, and unanticipated consequences.

Even when the policies *were* sound, Carter's approach left him with little interest in mastering the details of execution. Apart from a few areas, mainly in foreign policy, a president accomplishes very little simply by making the right decision; he must then convince the public, the Congress, or the permanent government if he wants to see his policies carried out. Carter took office without extensive background in any of these areas, and his disinclination to squeeze his employees for every bit of knowledge they possessed about doing these jobs meant that one of his initiatives after another died soon after it was announced.

There has been a good side to the kind of relationship Carter had with his staff: By keeping everyone in his own little niche and by not encouraging his economic advisor to criticize a proposal of his foreign affairs advisor, as, say, FDR would not have hesitated to do, Carter created a White House with less backbiting among the staff than any in modern memory.

Rivalry for the president's ear is the reason backbiting has been so common in other administrations. The Johnson White House saw many such rivalries; perhaps the most intense was between Marvin Watson and Bill Moyers. Under Nixon, Colson clashed with Haldeman and Ehrlichman; under Ford, Hartman with Cheney; and from the 1960 campaign until Kennedy's death, Richard Goodwin and Theodore Sorensen were rivals for the role of chief speech writer and chief brain, with Sorensen winning most of the time.

Sometimes the result of such rivalry can be healthy for the president and the country—as it was much of the time under Roosevelt—but this is seldom the case when the rivals are trying to out-do one another, not in force of argument, but in effort to please. At the court of King James I, the Catholic faction, finding the king between favorites and knowing his weakness for handsome young men, kept thrusting a lovely young male from their ranks across the King's path in the hope of gaining the most influence at court.

Under Lyndon Johnson, the technique appears to have been a bit duller. The basic method, according to George Reedy, who was Johnson's press secretary and later wrote a fine book, *The Twilight of the Presidency,* was "to be present either personally or by a proxy piece of paper when good news arrives and to be certain someone else is present when the news is bad."

Albert Speer has written of a court that, while more grotesque, was still similar to the White House in its basic behavior patterns: "The powerful men under Hitler were jealously watching each other. Bormann followed the simple principle of always remaining in the closest proximity to the source of all grace and favor. He accompanied Hitler to the Berghof and on trips, and in the Chancellery never left his side until Hitler went to bed in the early morning hours. . . ."

Jack Valenti took this same route to power under Lyndon Johnson. Such a relationship can deprive a person of his critical faculties. Remember when Valenti announced that he could sleep better at night because he knew Johnson was president? And consider this account of National Security Council meetings under Johnson, by Chester Cooper in his book, *The Lost Crusade*:

"The president, in due course, would announce his decision and then poll everyone in the room—council members, their assistants, and members of the White House and NSC staff. 'Mr. Secretary, do you agree with the decision?' 'Yes, Mr. President.' 'Mr. X, do you agree?' 'I agree, Mr. President.' During the process, I would frequently fall into Walter Mitty–like fantasy: When my turn came, I would rise to my feet, slowly look around the room, and then directly at the president, and say very quietly and emphatically, 'Mr. President, gentlemen, I most definitely do *not* agree.' But I was removed from my trance when I heard the president's voice saying, 'Mr. Cooper, do you agree?' And out would come, 'Yes, Mr. President, I agree.'"

If presidents suffer from too much heel-clicking obeisance from their staff, most presidents feel they could use a lot more from their cabinet and from the bureaucracy.

However, the basic rules of organizational life dictate that the departments and their secretaries will often not want to give the president what he thinks he needs.

The president's interest is in performance: Four years after his election, he must again go to the voters and be able to show that his government has done the job. The departments' interest is not performance but, of course, survival. Only a handful of appointees at the very top is likely to care about the survival of the administration as a whole, or to imagine that the president will have a clear enough view of what they're actually doing to tell the difference between cooperation and obstruction before his term is up. Even those who would theoretically like to cooperate are often persuaded by their subordinates and by their departments' constituencies to see the world more in terms of departmental interests.

A secretary of Labor may care more about the good opinion of the president of the AFL–CIO than he does about the president's, since simple political reality may dictate that the latter will have to keep a secretary who has the union leader's support. More common is the case of the agency head who, in order to win the loyalty of his subordinates, adopts their views as his own and champions their cause to the president.

Suppose the president determines that a drastic reorganization of the Department of Commerce to reduce the number of its programs is vital to the national interest. Naturally, this decision will be opposed by several individuals and groups in the bureaucracy: the Secretary of Commerce; the department's employees and their families; the congressmen whose committees oversee Commerce and whose domain would shrink as it shrinks; and, in all likelihood, the lobbyists who deal with Commerce and who usually prefer to keep the government operating in its old, ineffective ways rather than risk any change.

As each administration becomes aware of these facts of life in government, there is a tendency to exert ever stronger control from the White House, with final policies being worked out there instead of at the cabinet level. This tendency leads to an amusing spectacle, played out as a continuing drama as each new administration comes to town. The new president denounces the previous administration's centralization of power in the White House and says it will never happen again. At his press conference the day after the 1976 election, Carter said, "I would choose secretaries of Agriculture, the Treasury, of Defense, HEW, HUD, and others who are completely competent to run their own departments. I would not try to run these departments from the White House. The White House staff would be serving in a staff capacity—not in an administrative capacity." Then the old cycle began again. By the middle of 1979, newspapers were full of stories about how the Carter administration was firing cabinet members and reasserting authority over the executive branch. In 1981 Reagan began his administration with the same ringing endorsement of cabinet government, and the cycle was on its way to being repeated once again. But by July 18, 1982, the *Washington Post* was saying, "Rather than the corporate-style cabinet government promised by Reagan during the 1980 presidential campaign, most important decisions are made among the small group of White House advisers."

The administration immediately preceding Carter's had gone through a similar experience. When Gerald Ford became president, he announced that he would have no chief of staff. Instead, the lines of communication would radiate from him to his cabinet and senior staff like "the spokes of a wheel." Chaos was the result. Ford's time was frittered away in pointless meetings because no one was there to keep people from making appointments with the president merely to demonstrate their access to him.

Things changed, however. In 1975, he made Richard Cheney chief of staff. When Cheney left office he was presented with an object labeled, "The Spokes of the Wheel—A Rare Form of Artistry Conceived by Donald

Rumsfeld but modified by Dick Cheney." It was a bicycle wheel with all but one of the spokes twisted and disconnected. Cheney left the wheel for his successor, Hamilton Jordan, along with a note warning him to beware of the spokes of the wheel. But Jordan had to learn for himself.

The cabinet and the bureaucracy may be hard for the president to control, but they are as pliable as putty compared with the Congress. With few exceptions, such as the beginning of the New Deal [1933–1936], of the Great Society [1964–1966], and of Reaganomics [1981], congressmen have devoted themselves to thwarting the will of whomever happens to be residing at 1600 Pennsylvania Avenue, often without regard to whether he is a member of their party.

This normally difficult situation was not helped by Jimmy Carter's curious attitude toward the practice of politics. He seemed to regard this art with affection—even passion—during election years but with contempt the rest of the time. Once elected, he didn't want to bargain, to trade favors, to form alliances. He spent his first year in office not even knowing which key congressmen to call at which key moments.

Congressmen, like the president, need to be re-elected; and to get re-elected, again like the president, they must perform. But the kind of performance required of them is much different from what is demanded of the president: The president is asked to solve major national and world problems; congressmen are asked to solve individual problems their constituents have with the government. The president must bring groups together in a broad coalition to support his program; a congressman simply avoids offending them so they won't vote against him.

It is this fear of offending that makes a congressman especially vulnerable to the persuasion of lobbyists, who can threaten to withhold their groups' votes or campaign contributions—or worse, give them to his opponents. Thus it is through congressmen that lobbyists usually influence the president. The president has much less reason to fear the ten-vote loss that haunts the dreams of congressmen from districts where the total vote may be less than 1 percent of that cast in a presidential race. He may be willing to resist one group's special pleadings in order to serve the national interest. That's why the White House ranks sixth on the lobbyist's target list. But if that group has gotten to, say, a committee chairman who controls the fate of legislation important to the president, the president listens to that group.

But this also suggests the best way for a president to deal with Congress. And that is to make himself the most effective lobbyist of all. There are two ways to do this, the FDR way and the Lyndon Johnson way.

The FDR way was to mobilize the public through such devices as the famous Fireside Chats. Roosevelt spoke to the people; the people then spoke to Congress. In the period from 1933 to 1936, only a very brave congressman—or one from a safe Republican district—dared to oppose FDR, and an unprecedented amount of legislation was enacted at the president's urging. Fear of FDR's wrath (and, consequently, of losing the public's support and votes) kept Congress in line until Roosevelt first angered a large segment of the public by trying to pack the Supreme Court with his supporters and then attempted to purge the congressmen who had opposed him. The effort was a total flop, and FDR never again had such an easy time with Congress.

Reagan's charm and speaking ability, which are comparable to FDR's, brought him success for his programs in 1981. But his grip on the public began to slip as the people began to realize that his grip on the issues was, unlike FDR's, slim indeed. And by October 1982, Robert Shogan, the excellent political reporter for the *Los Angeles Times,* thus was publishing a book subtitled in part, "Why Presidents Fail."

LBJ, on the other hand, was incapable of mobilizing the

masses. His formal speeches were dull beyond belief. He was, however, a master at one-on-one cajolery and tireless in its practice. Few congressmen could resist his combination of passionate persuasion, down-home humor, and not-so-subtle arm-twisting, so the period from 1964 to 1966 is probably the second most productive in legislative history. His success came to an end for much the same reason as FDR's. He alienated much of the public with his policy of escalation in Vietnam, and many of his congressional supporters were defeated in the 1966 election.

Alexander Hamilton predicted in *The Federalist* that the office of the president "would amount to nothing more than the supreme command and direction of the military forces." The domestic power of the president has grown since those days, but it is still far less than his power in the national security area. An interesting illustration of just how weak a president can be on the domestic front comes from the later years of the Johnson presidency, after LBJ had lost his clout with Congress. In August 1967, Johnson decided that federally owned land in the cities should be used to solve the housing problem. His idea was to get the cities to build "new towns" within their limits using federal, state, and local money left over from existing programs. In this way, Congress—which in its prevailing anti-Johnson mood probably would not have consented—wouldn't even have to be consulted.

But he still had to deal with local mayors and city councils, who proved to be just as recalcitrant as the Congress. To top it off, the General Services Administration refused to give away the land without congressional authorization. So not only did he have to persuade the local governments to build the homes, he also had to persuade them to buy the land. They refused to do either.

At the same time, Johnson, as commander-in-chief, had 500,000 men carrying out extensive search-and-destroy operations in Vietnam. In that case he had power—he could push buttons and something happened. On the home front, the pushed button produced nothing but a flurry of activity by his own staff.

The fact that the button labeled "defense" or "national security" *does* produce results is a mixed blessing, as we have learned through painful experience in recent years. From the Ellsberg break-in to the Bay of Pigs incident to the Vietnam war, excess after excess has been committed in the name of national security. Even in the case of Watergate, generally thought to be a purely domestic tragedy, the smoking gun of the cover-up was Nixon's mis-use of national security, his order to Haldeman to have the CIA tell the FBI to stop investigating because national security was involved.

Pressure to cover up bad news, to avoid "leaks" of truths unpleasant to the president has been a constant source of evil in the culture of the White House. Even in the case of the Bay of Pigs, where Kennedy bravely accepted complete responsibility after the event, the cover-up mentality had a lot to do with how the failure happened in the first place.

The impression given by histories of the event is that Kennedy actually thought the invasion was a bad idea, but he also thought that cancelling it would show weakness. So rather than facing the issue squarely, he waffled and postponed. And rather than debating the pluses and minuses of the invasion with people who were likely to see the latter, he let himself be bamboozled by CIA director Allen Dulles and his deputy, Richard Bissell, who saw only the pluses. They also played on Kennedy's fear that a cancellation could not be covered up.

There were two important meetings about the Bay of Pigs in the spring of 1961. In the first, the planners of the operation dominated, and Dulles skillfully exacerbated Kennedy's fear of exposure by warning him that if he called off the operation it would mean the already trained guerrillas would be "wandering around the country telling the country what they have been doing." In the second meeting, Kennedy assembled a more reliable crew, but he made sure that he wouldn't get any useful advice from them by going around the room and demanding from everyone present a simple go or no-go opinion. The two men high in the administration who were not obsessed with the fear of appearing soft, Adlai Stevenson and Chester Bowles, pointedly had no role in the deliberations. Stevenson was kept in the dark throughout, and Bowles' memorandum opposing the invasion never got past Dean Rusk's desk.

When Kennedy finally made the crucial decisions about the invasion, he did so in solitude and gave his orders over the phone. Alone in his office, he called Bissell and told him to cut back the planned level of air support for the operation; alone again, he called Bissell and told him to start the invasion; and alone at his country house, he called Rusk and canceled a second air strike at the landing site, thereby completely assuring the operation's failure.

The only good that came out of the fiasco was that Kennedy learned some lessons that helped him deal more successfully with the biggest crisis of his administration. By the time he discovered in October 1962 that the Russians were secretly installing missiles in Cuba, he had effectively communicated to the national security bureaucracy that he did not reward those who lied to him. Dulles was out; so was Bissell. And this time Kennedy cross-examined each of his advisors to bring out any hidden assumptions or subtle doubts. He also broadened the advisory group to include intelligent non-experts like his brother.

And Robert Kennedy's role in the Cuban missile crisis turned out to be crucial. He was given the most sensitive task of the entire negotiations, letting Soviet Ambassador Anatoly Dobrynin know that we were going to pull our missiles out of Turkey. This gave Khrushchev the *quid* he needed for all the *quo* he was having to swallow. But John Kennedy was so anxious to *appear* tough that he never let the American people know about the assurance to Dobrynin—indeed, he denounced Adlai Stevenson for suggesting such a course.

However, if Kennedy had not learned to admit he had a peaceful side, he had learned other lessons of the Bay of Pigs. Jimmy Carter, on the other hand, was slower to profit from his own Bay of Pigs, the energy program. It had been sold to him by James Schlesinger, who, like Dulles, made sure that potential critics from within the government were not fully informed or informed too late to mount effective opposition. And Carter, like Kennedy, failed to involve advisors who were likely to find fault with the program.

But unlike Kennedy, who fired Dulles within six months, Carter kept Schlesinger for two and a half years and continued to fail to involve people outside the regular chain of command in the decision-making process, to encourage argument among his advisors, or to encourage them to argue with him, except on paper, where there is no give and take and much less chance of real impact. And Carter again failed to act when his military planners produced a disaster at Desert One in Iran.

Recent history presents each president with an agenda. There are problems on his desk when he arrives, like the Mid-East crisis that was waiting for each of the last four presidents. Each day's events produce new problems that must be faced and ceremonial events, from laying a wreath to greeting a prime minister, in which he feels obliged to participate. And every day there are the memoranda, the telephone calls, and the meetings that tell him what the rest of the government thinks should be his agenda.

The problem is that we need a president who refuses to be ruled by the agendas of others. We need a president who has a program of his own, who looks beyond the problems that are obvious today to those that could confound us tomorrow.

But having a program is itself not enough. We have to take a hard look at how sound it is, at how much sense it makes. After all, Ronald Reagan had a program, but it just happened to be the wrong program. And the result was severe economic distress for the nation.

In addition to looking at the program a presidential candidate offers, we should look at the character of the man behind it. Character is the ability to rise above all the forces that keep us from thinking clearly—not only about what will work but about what is right. Johnson and Nixon, even Ford during the Mayaguez incident, displayed a crucial lack of it in yielding to the inner need to appear tough that kept them from seeing the right course in Southeast Asia.

Kennedy turned the failure of the Bay of Pigs into the success of the missile crisis by overcoming his fear of being soft in time to tell the Russians he would remove our missiles from Turkey. But if only he had not been afraid of letting the public know what he had done! The tragic result of this fear was that people saw the missile crisis as a triumph of toughness and saw the lesson as a need to demonstrate our "resolve" in Vietnam. One wonders about the lives that could have been saved if Kennedy had told the truth about his dovish gesture. Maybe then he could have told the public what he in fact told only two men: that he thought Vietnam was a mistake and planned to pull out after the 1964 election.

Robert Kennedy finally revealed his visit to Dobrynin in a book published shortly after his assassination in 1968. But this was over five years after the event—and three years and thousands of deaths after we demonstrated our resolve in Vietnam by the escalation of 1965. It is hard to learn from history if your leaders hide it from you.

A WAY OUT

The title of this book may be misleading. The truth is that Washington doesn't really work. The government is not solving the nation's problems. The question, of course, is what, if anything, can be done. My answer is simple: Let's try democracy. Far too many of the decisions that govern our lives are made by bureaucrats and lobbyists who are not accountable to the people.

The bureaucrats aren't accountable to the people because they can't be fired by the elected representatives of the people. If elections are going to mean anything—and they often mean nothing now because they do not change the permanent government—the administration must be given the authority to hire and fire not just cabinet members and agency heads, but enough other officials, high and low, to allow the president to move the machinery of government. This power would also give presidents the patronage needed to rebuild the political parties as bulwarks against the threat of single-issue politics. Power over federal appointments within a congressional district should be shared with the congressman—just as administrations have for years shared with senators the power over the appointment of federal judges and United States attorneys—or with local party officials when the congressman is not in the president's party. That way the voters will know that the congressman shares responsibility for the performance of federal officials in his district. Then, if your postman spends his time reading his mail instead of delivering yours, you will know where the blame belongs.

If we are going to truly reform the American system of government, we need more politics rather than less—more good people running for office and helping other good people run for office. If you never again want to face a choice between a Jimmy Carter and a Ronald Reagan, you must make sure you have a lot more people to choose from, a lot more good people pursuing political careers.

This in turn means offering decent rewards for a life in politics. Today a person who starts out in politics has a very tiny field of opportunity—congressman, senator, governor, and just 2,000 federal appointive positions. But what if we opened up hundreds of thousands of federal jobs to political appointees, replacing as jobs come open through normal attrition roughly 50 percent of the federal government's 2.8 million civilian employees? Give the new people two-and-a-half-year appointments, with a limit of five years on the time they would be permitted to remain in government. This would:

1) bring people with real world experience into government

2) attract a different kind of person, the risk-taker, who is not interested in job security and who would be less cautious and self-protective than civil servants

3) provide a legitimate reward for political participation and thus attract more people to political activity. (The reward would be legitimate because the unqualified would not profit from it. Your sister Susie who can't type 20 words a minute will not get that government typing job no matter how hard she worked on your campaign.)

4) send back to the nation as members of the voting public literally millions of people who have personal knowledge of how the government doesn't work, of the reforms that are needed to make it function, and of how to bring them about.

But suppose the Reaganites were permitted to begin

making such appointments now? They would do some harm. That I cannot dispute. But you could elect a good president in 1984 and replace them all in 1985 with people who are dedicated to the program of the new president. Because you could see the difference they would make, I think you would be much more concerned with the president you elect. Remember, the last president who did great things was FDR. And he hired, not the two thousand employees recent presidents have been able to choose, but two and a half *million.* We don't have to go that far, but it does suggest the direction in which we need to travel.

All this is not to say that I think the career civil servant serves no useful purpose. If I felt that way, I wouldn't advocate retaining half of them as I do. They provide continuity, institutional memory, and an insurance policy against the excesses of the politicians. All of these are valuable roles. But we also need people with other virtues, the kind of people short-term appointments would attract. Most of all we need to provide the incentives that will attract people who can revitalize our political system. If we don't want a system that runs on money, then we have to offer something else. What is better to offer to the people who push the doorbells and hand out the leaflets than the opportunity to participate in putting into effect the programs they have campaigned for?

If this program had been in effect for even a few years, we now would have a nation far better equipped to make the budget cuts that are needed. We would have a public that didn't have to guess where the fat was because they'd know exactly where to find it.

Finally, we would have people in government who, because they've spent most of their lives on the outside, would have genuine empathy for the problems of those on the outside. The lack of such empathy has been the most glaring deficiency of the bureaucracy since I have been in Washington. And I fear it will become worse, because the mindless Reaganite attack on the bureaucracy is going to exacerbate the civil servants' tendency toward self-protection just as did the equally mindless McCarthy attack of the fifties.

In *Elrod vs. Burns,* and again in *Branti vs. Finkel,* the Supreme Court said that public officials below the policy-making level can't be fired on political grounds because that would penalize them for their ideas. The justices can't have been serious—and will, I'm confident, reverse themselves when the first lawyer pierces the fog of civil service respectability and explains to them that executing policy is as important as making it, and that the point of democracy is to let the voters remove officials whose ideas they don't like.

With an accountable bureaucracy, congressmen will no longer be justified in spending 80 percent of their time on constituent service. They can reclaim the legislative functions that have gone to the regulatory agencies and the courts by delegation or default. They can do a proper job of overseeing the activities of the executive branch. They and the president, because they share the appointing authority for the federal officials working at the grassroots level, should finally be able to overcome the crucial defect in the government's information system—the communication gap between the people at the bottom and those at the top, the gap that means those at the top seldom know the truth about what's going on down below and therefore don't know what law to pass or what orders to give.

It is equally important that this information reach those on the outside, that the people find out what is going on at both the top and the bottom—whether it's what John Kennedy did about the Turkish missiles or whether that missing social security check is an isolated mistake or part of a larger problem their government should do something about.

The main responsibility for providing this information belongs to the press. To fulfill its role, the press needs to develop a new breed of reporters who know the history of the American political system and the lessons of its successes and failures—who have the descriptive skill to report accurately what the government is doing, the analytical skill to figure out how what it is doing can be done better (or whether it should be done at all), and the literary skill to make the description and analysis something people will read and remember. I don't mean that every reporter must combine the best qualities of Samuel Eliot Morison, Seymour Hersh, Benjamin Cardozo, and Vladimir Nabokov. I do mean that reporters and their employers must raise and broaden their aspirations so that success on our great papers comes to mean not only rooting out a scandal and getting a politician indicted, but telling us what's wrong with the system in a way that also tells us how it can be made right.

None of this will make any difference, however, until a majority of us, reporters included, are willing to renounce our own roles in the game of make believe. This will be hard for the reporter who wants to believe that covering those presidential trips is important. It will be hard for nearly everyone, in fact, because most of us have a stake in make believe. The congressmen who only pretend to consider genuine reform of the tax law or serious reduction of government spending reflect our own reluctance to jeopardize the loophole or expenditure that benefits us. The main reason the lobbies usually win is that each of them represents at least some of us.

What is wrong with Washington, then, is what is wrong with the rest of us. It won't be cured until we have a rebirth of patriotism—or, if that word embarrasses you, a willingness to put the welfare of the national community above our own. Maybe it won't happen. But it can. The proof is as recent as the 1980 Iowa caucuses. Two weeks before they took place, President Carter announced a grain embargo against Russia. The embargo threatened the Iowa farmer and was denounced by all but one of the other Democratic and Republican candidates. It was expected that Carter would suffer politically. Instead he won by a wide margin. If the Iowa farmer can do it, why can't the rest of us?

INDEX